Fungal Reproduction and Growth

Edited by Sadia Sultan
and Gurmeet Kaur Surindar Singh

Published in London, United Kingdom

IntechOpen

Supporting open minds since 2005

Fungal Reproduction and Growth
http://dx.doi.org/10.5772/intechopen.95215
Edited by Sadia Sultan and Gurmeet Kaur Surindar Singh

Contributors
Kamal Prasad, Agam Khare, Prateek Rawat, Ozlem Gulmez, Ozlem Baris, Chinaza Godswill Awuchi, Erick Nyakundi Ondari, Ifie Josiah Eseoghene, Hannington Twinomuhwezi, Ikechukwu Otuosorochi Amagwula, Sonia Morya, Vasanthabharathi Venkataraman, Jayalakshmi Singaram, Kalaiselvi Vaithi, Faisal Rasheed Anjum, Sidra Anam, Muhammad Luqman, Sajjad ur Rahman, Muhammad Sohail Sajid, Farzana Rizvi, Ameena A. AL-surhanee, Abdullah F. F Shater, Muhammad Wasim Usmani, Muhammad Zulqarnain Shakir, Manish Mathur, Neha Mathur, Khirood Doley, Mahesh Borde, Susan Thomas

First published in London, United Kingdom, 2022 by IntechOpen
IntechOpen is the global imprint of INTECHOPEN LIMITED, registered in England and Wales, registration number: 11086078, 5 Princes Gate Court, London, SW7 2QJ, United Kingdom
Printed in Croatia

British Library Cataloguing-in-Publication Data
A catalogue record for this book is available from the British Library

Additional hard and PDF copies can be obtained from orders@intechopen.com

Fungal Reproduction and Growth
Edited by Sadia Sultan and Gurmeet Kaur Surindar Singh
p. cm.
Print ISBN 978-1-83969-878-1
Online ISBN 978-1-83969-879-8
eBook (PDF) ISBN 978-1-83969-880-4

For EU product safety concerns:
IN TECH d.o.o., Prolaz Marije Krucifikse Kozulić 3, 51000 Rijeka, Croatia, info@intechopen.com or visit our website at intechopen.com.

Meet the editors

Sadia Sultan received a Ph.D. in 2004, following which she worked as a chemist in the QA Department of Abbot Lab, Pakistan. In 2006, she was appointed as a lecturer in the Faculty of Pharmacy, UiTM Shah Alam, and associate research fellow at Atta-ur-Rahman Institute (AuRIns) for Natural Product Discovery previously known as IKUS. She has 18 years of experience in the field of natural product research (NPR). Currently, she is an associate professor in the Department of Pharmaceutical Chemistry and Pharmacology. She is the author of several books and book chapters, and she has more than sixty research publications in top international journals to her credit. She has also edited one book. She has been appointed as an organizing committee member for several international conferences. She has acquired numerous research grants from the Ministry of Higher Education (MoHE), Malaysia, as a main and co-investigator since 2006. Her current research is based on biotransformation and isolation and characterization of fungal secondary metabolites using modern NMR methods. She is a journal reviewer and an editorial board member of the *Open Enzyme Inhibition Journal*, *Frontiers in Pharmacology*, *Oriental Journal of Chemistry*, and others. Her recent research focuses on dereplication using LC/MS/MS and capillary probe NMR. This is a new area whereby screening on a microscale is done via automated systems and databases.

Dr. Gurmeet Kaur Surindar Singh is a senior lecturer and the deputy dean of Research & Innovation, Faculty of Pharmacy, Universiti Teknologi MARA (UiTM), Malaysia. She completed her Ph.D. in Medicine at the University of Sydney, Australia, in 2015. Her current research interests include neurosteroid metabolism in Alzheimer's disease (AD), regulation of phase II enzymes and neurosteroids in AD, steroid analysis in biological samples using liquid chromatography-tandem mass spectrometry (LC-MS/MS) (method development and validation), neurophysiology and neuroendocrinology, the reproductive study of aphrodisiac plants, and neuroprotective role of astaxanthin nanoemulsion in AD. She has published several papers in reputed endocrinology journals, authored one book chapter, and acquired numerous fundamental research grants from the Ministry of Higher Education (MoHE), Malaysia, as a main and co-investigator since 2010. She has presented her research findings at international conferences such as the International Congress of Endocrinology, Chicago, and the American Society of Mass Spectrometry, Minnesota. She has also won gold awards for invention competitions. Currently, Dr. Gurmeet is the editor-in-chief of the *International Journal of Pharmaceuticals, Nutraceuticals, and Cosmetic Science* (IJPNaCS).

Contents

Preface

Fungi are eukaryotes that present as pathogens, parasites, and symbionts in the ecosystem. There are more than tens of thousands of fungal species described. Chapter 1 of this book provides an overview of fungi reproduction, pathology, and the unique features that allow fungi to survive in diverse ecosystems. Chapter 2 discusses fungal mycotoxicity and pathogenicity and the suitable environmental conditions necessary for fungi to thrive. In favorable conditions, fungi produce mycotoxins that pose a challenge to the agriculture industry, hence prevention and control measures are needed. Fungi are also found in marine habitats, albeit in low numbers. Marine fungi are the key component of the sponges' natural diet. Chapter 3 examines the development and growth demand of fungi. This study enables a greater understanding of fungi and how to take precautions against these organisms. Chapter 4 describes how fungi are detected by the host's major pattern recognition receptors and the mechanisms that fungi use to escape the host's immune response. The majority of fungi are non-motile; therefore, it is important to explore the evolution of fungi mating mechanisms. Chapter 5 discusses fungal hormones/pheromones such as sirenin, trisporic acid, antheridiol, oogoniol, and peptide hormones. Chapter 6 examines how arbuscular mycorrhizae facilitate host plants to grow vigorously under stressful conditions. The final chapter deals with several abiotic and biotic factors that affect the growth of keratinophilic fungus and the substrates that could potentially function as growth promoters in this process. Finally, Chapter 7 explores marine sponge-associated fungi and their secondary metabolites as therapeutic agents depending on their bioactive properties. However, these fungi species are also associated with a wide spectrum of diseases, including HIV, cancer, and others.

This book provides readers with an in-depth understanding of fungi diversity and the role of fungi in the ecosystem.

Sadia Sultan and Gurmeet Kaur Surindar Singh
Faculty of Pharmacy (UiTM),
Bandar Puncak Alam, Kuala Lumpur, Malaysia

Section 1

Fungal Growth

Chapter 1

Abiotic and Biotic Factors: Effecting the Growth of Keratinophilic Fungi

Manish Mathur and Neha Mathur

Abstract

Fungi portray an important role in decomposition of keratin, as their activity is tough to measure. According to an estimation, a quantity of cellulose is synthesized by primary producers over photosynthesis and then reinstated to the atmosphere as carbon dioxide and through the activity of fungi, which decompose the complex and inflexible polymer. Without this activity, the world would soon be submerged by plant residues, and this would probably exclude most living organisms from their natural habitat. This chapter deals with several abiotic and biotic factors, which effect the growth of keratinophilic fungus and the substrates, which can serve as potential growth promoters for them.

Keywords: keratinophilic fungi, growth factors, fungus substrate

1. Introduction

The degradation is associated to the aptitude of the organism to develop on a substrate. The capacity of the growth depends upon the enzymes secreted by the microbe and conditions such as PH, protein content in the substrate, temperature, moisture, etc. The native feather keratin was degraded by species of *Aphanoascus fulvescens* and *Chrysosporium articulatum*, which are separated from soil. Recognition of strain was done by phenotypical qualities and nucleotide sequencing of protein. The ultimate deficit of substrate in comparison to the strain was recorded for *Aphanoascus fulvescens* and deficit of substrate for the *Aphanoascus fulvescens* (71.08%). Keratin is high in protein and abundant in nitrogen and sulfur [1–3]. Keratins have high mechanical and chemical resistance. Numerous disulfide bonds (S–S) are liable for the resistance. Except a few months, only keratinolytic microorganisms can grow and degrade native keratin including some bacteria of the genera *Bacillus, Vibrio, Serratia* ([4], actinomycetes [5]). The genus *Bacillus* [6], representative geophilic dermatophytes are associated with the fungi known as *Chrysosporia*, name taken from the genus *Chrysosporium* [7]. The development of fungi is backed by a humus, neutral or a little alkaline pH reaction, fruitfulness in $CaCO_3$ [8]. The *Chrysosporim* belongs to this group, specialized in the growth of keratin, e.g., feathers, hairs [9]. Although it is considered safe in certain stages but can be converted into pathogenic form under certain conditions. Kushwaha [9] reported that genus *Chrysosporium* is effective in disorienting keratin; however, keratin degradation varies in the species. Some microbes grow on keratinous materials including bacteria, biodegradation has been exhibited by *Bacillus*, particularly *Bacillus licheniformis* [10] and *Bacillus subtilis* [11], and *Chryseobacterium*. Actinomycetes from *Streptomyces* genus produce keratinases [12, 13]. The most common keratinolytic fungi are *Aspergillus*, *Penicillium* [14], *Fusarium* [15].

2. Growth factors in keratinophilic fungi

To-Ka-Va hair baiting technique [16] was used in the experiments.
Several workers projected that saprophytic phase of dermatophytic fungi in the soil are global [17, 18]. Dispersal of pathogenic fungi [17, 19–22] specifies soil turns as pool meant for primary contamination may be for some pathogenic fungi. Potentially it is pathogenic to wild animals and acts as the resource of secondary infection to man and pets [23–26]. Besides soil, birds' nests and feathers [27–35], hairs [25, 36], water [37], plant debris [38], and dung [39] are the different ecological places for growth of capable dermatophytic fungi.

Fungi produce enzymes potent of degrading keratinized forms known as keratinophyles. Pathogenic strains find a suitable host in favorable environmental and physiological conditions and produce the symptoms known as "ringworm diseases" by growing on human dermis.

2.1 Climatic factors

2.1.1 Temperature

Keratinophilic fungi are mesophilic while a few have grown at 37°C such as *C. tropicum, Chrysosporium keratinophilum, C. queenslandicum*, etc. Nevertheless, geophilic dermatophytes grow best at 25–30°C. In 1970, Pugh and Evans mentioned span of 25–27°C in the keratinophilic fungi, which do not grow more than at 40°C.

The broadly recognized view about mesospheric in nature has been disquieted by the results of Battelli et al. [40], about the existence of *Microsporuzri pypseiim, Trichophyton ajelloi, T. terrestre* in alpine mountain. Pugh and Allsopp [41] stated a recurring existence of *Chrysosporium pannorum* and *Mortierella* spp. with exceptional incidence of *Trichophyton terrestre* and *Chrysosporium* sp. Some varieties are thermotolerant, and recently, it was shown that they can adapt to adverse temperatures for survival [41].

2.1.2 Light

It was known that the UV light is fungicidal, which creates about a reserve of spore sprouting leading to a hypha. Excellence of illumination on spore germination has entirely been analyzed by Buchnicek [42].

First shown by Berde [43], the growth reticence of dermatophytes is via means of lamp. Fungicidal effect in *Candida albicans* differed with the strength of light sensitizer and light intensity as was observed by Dickey [44]. There is no effect on illumination of spore suspension of *T. mentagrophytes* without photosensitization [44]. The red or blue light when applied separately prevents the growth greater than visible light. The quantity of inhibitions by both color lights utilized individually is more than the reserve by their mixture.

2.1.3 Seasonal variations

Season is quite anticipated with respect to dermatophytic flora of the soil, which has climatic differences resulting in change of temperature. *Keratinomyces ajelloi,*

Trichophyton Terrestre, and Microsporum gypseum were found in soil samples during April and in August.

2.1.4 Soil pH

Effects of pH on microbial life are explored extensively. Bohme and Ziegler [45] observed about soil pH as verified by Pugh [32]. Ziegler [46] observed the degradation of keratin over a wide range. Findings uncovered the optimum pH for amylase 7–2, alkyl phosphates 8–7, lipids 6–8, proteinase 5.4–6.9. Ectoenzymes are shown dormant at pH below 4.5, and the enzymatic catabolism of keratinophyles occurs at pH 6.9. Ziegler [46] further confirmed that the most luxuriant growth and highest frequency of keratinophilic fungi were seen in soils with pH 6.9.

2.1.5 Carbon

Decomposition of carbon can be done by keratinophilic fungi but there is not enough information regarding carbonate impacting the delivery in soils. Chmel et al. [47] described favored incidence of *M. qypseum* in carbonate field soil, an alliance of *K. ajelloi* with soils and a predilection of T. *terrestre* in alluvial soils. It was studied that a plenty of keratinophilic fungi in carbonate field and chernozemic soils have greater humus substance in comparison to gray podzolic soils. Keratinophyles can decompose extremely compound carbon complexes in many culture media, a simple carbon source can be used and not influence their qualitative spreading but the findings on their quantitative dispersal may disclose fascinating truths.

2.1.6 Nitrogen

Keratinolytic fungi can decompose keratin, which is a rich source of nitrogen. Rate of keratinolysis has been reported by Ziegler [46] at pH 6–9 supporting the earlier finding.

Growth of *Microsporum gypseum* is fine revealed by the experiments utilizing optically dissimilar forms of cystine, quite large amounts in the scleroprotein [48]. Carbon and nitrogen application was perceived when surplus sulfur was secreted in the medium and became oxidized. Oxidation of sulfur in the extracellular medium makes use of L-cystine, which was showing a slow form. The way of consumption of L- and D-L-cystine was the same. The nitrogen-containing substances of keratin differ from one another and in opportunity would impact colonization of the last.

2.1.7 Sulfur

Kunert [49, 50] calculated that the inorganic and organic sulfur resources for the development of the fungus *M. gypseum.*, sulfite, disulfide, peroxodisulfate dithionate, and sodium sulfate were the best hints of inorganic sulfur. Inorganic sources sulfide produced a primary inhibition of the fungal growth supportive to a previous report that mineral water comprising H_2S is repressing for the growth of dermatophytes. Amino acids such as cystine, cysteine, glutathione, S-sulfocysteine, lanthioneine, taurine, and serine-sulfate are the leading organic sulfur sources of the fungus.

2.1.8 Moisture content

An enzymatic reaction takes place in aqueous solution in the cell cytoplasm and is vital to life. The systems of germination, growth, and reproduction are energetic associated to the substrate moisture content. Pugh and Evans described a better percentage of spore growing in *Arthroderma uncinatum* and *Ctenomyces serratus* at 90–100% RH. It was also stated that an plenty of keratinophyles in birds' nest with moisture usually showed a greater occurrence of keratinophilic fungi in bird nests with 15–20% water content [51].

2.1.9 Humus

High humus content in soils invites M. *gypseum* and *T. terrestre* in soils as revealed by Chmel *et al.* [47]. Whereas distribution of K. *ajelloi* and *C. keratinophilum* is regardless of soil humus. Soils made up of fragmented lava with little organic matter in the Galapagos Islands revealed by Ajello and Padhye [52] reported the occurrence of *A. quadrifidum, C. indicum*, C. *keratinophilum, C. tropicum,* and *Ctenomyces serratus*. A higher number of keratinophilic fungi were described in soils with exceptional humus value. In all levels of soil profile extreme in the region with high humus content, there exist *M. gypseum and T. georgii,* while *T. oanbreuseqhemii* and its flawless state *A. gertleri* were primarily found in soils with low humus value.

2.1.10 Fatty acids and oils

In 1899, Clarke antifungal estates of fatty acids were known to scientific world. The anti-dermatophytic properties were discussed by Rigler and Greathouse in 1940, which was later validated by Das and Banerjee [53].

Hajini et al. [54] studied unsaturated fatty acids, hair oils, and various natural fatty acids for their anti-dermatophytic properties. Reticence of growth of *T. rubrum* and associated dermatophytes was at 0.1% strength of mustard oil, oleic, linoleic. Linolenic, and aracliidionic acids while coconut oil, castor oil, till oil, Bryl cream, vaseline hair tonic, palmitic acid, and stearic acid did not prevent the growth even at 10% strength.

2.1.11 Salts

In coastal soils, various keratinophilic fungi can survive. *A. curreii* occur although the rate of availability of keratin must be small. Varieties of hares, rabbits, and birds exhibited *A. curreyi* and *C tenomices serratus*. Padhye *et al.* [55] isolated *Clirysosporium tropicum* and *Microsporum* gypseum from long immersed marine soils in Bombay. *C. indicum* and *Cteiiomyces sermons* and inaccessible to somewhat irregularly. The fungi remain to be revealed, for no such fungi have been found from coastal Mediterranean soils, which have salinity. A repressive sodium chloride impact on the growth has been stated of dermatophytes. Growth was reserved by NaCl in *Microsporum, Epidermophyton, Trichophyton,* etc.

2.2 Biotic factors

Biotic component influences the occurrence of keratinophilic fungi as the main causal factor in spread and persistence of ringworm diseases produced by dermatophytes up to a great extent. Biotic component influences the occurrence of fungi in birds and animals.

2.2.1 Birds

Pugh [56] displayed the existence of keratinophilic fungi on the experimental birds although validated them on birds in Australia. An association exists in the keratinophilic fungi and birds of correct order, *e.g., A. curreyi* and *Turdus* [31]; *C. serratus* and representatives of Galliforme, particularly partridges and chickens. Regularly conveyed from most of the bird forms [56], in *Chrysosporium* spp. Commonly found species are *C. k eratinophilum, Keratinomyces ajelloi,* and *T. terreste*. There was significant habitation for the cleistrocarpic stage of *C. serraius.*

2.2.2 Animals

The role of wild animals as carrier of these diseases has been documented earlier in literature.

Occurrence of *Tinea capitis* was in newborns who obtained disease from stray kittens. Incidence of scalp abrasions in an 8-year-old boy and a 3-year-old Yorkshire was seen as well as subsequent isolation of *Microctenopoma nanum* from soil. Soil acts as a group for dermatophytes [57].

3. Fungal growth on feathers

Keratin makes feather obstinate to common proteases such as trypsin, pepsin, papain, slowing its degradation process. Each bird has up to 125 gm of feather and approx. 400 million chickens processed universally the daily accumulation of which reaches 5 million tons [58, 59]. According to Lin *et al.* [60], the waste disposal is a global issue foremost to pollution of both air and water resources. Keratinase-treated feather is considered as a viable source of dietary protein in food and feed, as it has high nutritive value. Keratinases are potential market as proteases. Microorganisms are described to produce keratinase as *Doratomyces microsporus, Alternaria radicina, Trichurus spiralis, Aspergillus sp., Rhizomucor sp., Absidia sp,* etc., and *actinomycetes as Streptomycespactum, S. alvs, Streptomyces thermoviolaceus, Streptomyces fradiae, Thermoactinomyces candidus* etc.), and as bacterial species (*Fervidobacterium islandicum, Pseudomonas aeruginosa, Microbacterium sp.*, and *Bacillus* including *Bacilluslicheni formis* and *B. pumilus*) [61]. Feather is insoluble fibrous protein and highly resistant [62] to enzymatic digestion [63]. Though, fungi often colonize on various keratinous substrates, degrade them and enhance the minerals in soil [64].

Hydrolytic enzymes are synthesized by filamentous fungi. Various species are used to produce industrially important enzymes as distinct proteases, carbohydrates, and lipases. It is the key enzymes in fungal incursion of skin found in dermatophytes as *Trichophyton* [65]. *Candida* also contributes to skin infections. Enzymes are found in *Streptomyces* [66] and *Bacillus* spp. [60]. Novel and Nickerson [67] examined bacteria, actinomycetes, and fungi for keratinolytic activity and found *Streptomyces* as most active in the decomposition of sheep wool. Keratin hydrolysis was the most active in *Verticillium tenuipes, Trichophyton equinum,* and *T. mentagrophytes* in peacock feathers. *T. mentagrophytes, T. verrucosum,* and *Keratinomyces ajelloi* degraded hair, whereas only *T. gallinae* degraded chicken feathers.

4. Fungal growth on hair

Keratin, the fibrous protein, is a codified part of hair, wool, and related structures, which differ from other proteins in their high cystine content.

Five keratinophilic fungi, i.e., *Chrysosporium indicum, Geotrichum candidum, Gymnoascoideus petalosporus, Scopulariopsis brevicaulis,* and *Talaromyces trachyspermus,* which grow on human hair in stationary culture, have been examined. Hair was studied on criteria of cysteine, cystine, inorganic sulfate, thiosulfate, total protein, keratinase, and change in alkalinity. *Gymnoascoideus petalosporus* showed degradation to remaining isolates when grown on human scalp hair as the sole resource of nutrients *in vitro.*

Kunert [68, 69] described a release of cystine, cysteine, and sulfate in the culture filtrate of *Microsporum gypseum* growing on hair. Ruffin et al. [70] discovered the S-sulfo cysteine in culture fluid and established the role of sulfitolysis during keratin degradation by *Keratinomyces ajelloi.* Stahl et al. [71], Chesters and Mathison [72], and Ziegler & Bohme [73] could not detect cysteine in filtrates of the dermatophytes studied. Weary et al. [63] recorded the production of 21 pg./ml and 38 pg./ml of cysteine by two strains of *Trichophyton rubrum.*

5. Fungal growth on leather

Samples collected from different museums of feather and leather objects and deposited dusts were studied for the isolation of keratinopliilic and non-keratinophilic fungi. Throughout the study, five species of *Chrysosporium*, four of *Aspergillus*, one of *Penicillium,* and two each of *Acremonium* and *Fusarium* were isolated.

Deterioration of objects of cultural value, in the conservation of cultural heritage, is a real problem. Microbial activity on museum objects, especially skin or leather, starts with a surface infection that eventually invades the full thickness of the skin. The most common molds belonging to the genera *Aspergillus* and *Penicillium* emerge as black and green surface discoloration. There are, however, other molds, although of rare incident compared with the above, causes degradation of skin objects [74–76].

From different museums of Northern India, 24 keratinophilic and non-keratinophilic fungi represented by five genera and 15 species were isolated. Strains of *C. keraiinophilum,* three of *C. tropicum,* two each of *C. evolceanui* and *C. indicum,* and one *Chrysosporium* sp. were found in museums. Some non-keratinophilic fungi, i.e., three isolates of *Aspergillus flavus,* three of *A. niger,* one each of *A. sulphureus* and *A. luchuensis,* two species of *Penicillium, i.e., P. ciirinum, P. chrysogenum,* and two species of both of *Fusarium* and *Acremonium* were found and called as non-keratinophilic [77]. These were involved in the degradation of feather and leather objects.

Keratinophilic and non-keratinophilic fungi arise in regions where they can find dissimilar types of keratinaceous substrates. The existence of the fungi was connected with numbers of people inhabiting [47]. English [78] found 31 saprophytic fungi capable of inhabiting keratin along with non-keratinophilic fungi on keratinic substrate witnessed by Nigam and Kushwaha [77]. *Aspergillus* and *Penicillium* colonized keratinous substrates, and pathogenic behavior was revealed by Kishimoto and Baker [79]. The capacity of these fungi to colonize keratinous substrates was confirmed by Carmichael [80]. Some saprophytes are involved in microbial deterioration of leather [81, 82].

Keratinophilic fungi were commonly found on birds and animals and also isolated in the Antarctic [74]. Isolation of *Chrysosporium* spp. is supported by other work, and it was frequently isolated [55, 83–88]. The presence of *Chrysosporium* and related keratinophilic fungi was reported from museum objects, their surroundings, or their deposited dusts.

In 2006, several valued leather objects were found during archeological excavation of Ghalee-Kooh-i Ghaen (historic stronghold from the Seljuk period, 11th–13th centuries) in the South Khorasan province of Iran. When examined after 5 years, there were red stains on the fragments of a shoe with poor strength and powdery surface like red rot decay. Since red rot is more common in manmade leathers from the mid-nineteenth century as clarified by the structural features and degradation factors responsible for red stains on the shoe.

As leather production is an ancient industrial activity [89], historical and archeological leathers represent an important part of any society's cultural materials. *Ghalee-Kooh-i Ghaen is* a fortified castle from the Seljuk period (11th–13th centuries) located 3.4 km from Ghaen city in the South Khora-san province of Iran [90]. It was destroyed in 1066 A.D. by an intensive earthquake in the Ghaen area [91]. Later, Hossein Ghaeni rebuilt the castle in the late eleventh century to use as headquarters in the southern part of Khorasan known as Ghohestan [92]. The castle was registered as no. 4803 in the list of Iranian national monuments due to its historical and archeological value in 2002.

Several leather bottles, shoes, a fur, and some pieces of leather were found in the archeological investigation. Studies of the leather bottles indicated goat skin treated with lime depilation, vegetable tanning, and animal fats as lubricant for leather making. After excavation, the leathers were stored in inappropriate conditions at the base of the cultural heritage organization of South Khorasan province without any preventive measures. During the time of storage, there was no information about the environmental condition of the leathers within their archeological context in 2011 one of the excavated shoes was examined. Red rot is more common in manufactured leathers from the **mid-nineteenth** century, but the decomposition pattern may be found in more ancient leathers as well [93]. Therefore, this leather shoe was studied to better understand the deterioration mechanism prior to any interceptive activities.

Previous investigations on the structural features and degradation of leather and parchment artifacts have revealed important information, which is of value to our study [94, 95]. Sulfuric acid is regarded as main deterioration factor of red rot [96], possibly originating from environmental pollution or the materials used in the leather-making process [97]. Additionally, the production of acid from vegetable tannins, especially condensed and disintegration of the collagen-tannin complex, may result in red rot [96]. Many methods are available for identifying the decay, such as assessment of esthetic properties, pH and evaluation of the physical and mechanical characteristics of the object [96–98]. Red rot probably occurs in leathers with pH under 2.8. However, the degree of degradation cannot be determined just by pH measurement [98]. Moreover, biodeterioration is another important factor altering the artistic and useful properties of leather [99]. Fungi species attach to materials such as paper, textile, wood, paints, leather, etc., and produce characteristic signs that can be used for identification [100]. Leather composed of collagen, i.e., fibrous protein, provides a source for the evolution of proteolytic fungi. Some species of Aspergillus and *Penicillium* have been identified as the most common species related to leather and parchment molds [101]. There are also other rare molds responsible for biodeterioration of proteinous materials. Ebrahimi et al. [100] observed the activity of *Aspergillus* spp., *Penicillium* spp., *Chrysosporium* spp., *Madurella* spp., *Trichophyton* spp., and *Zygomycota* on the leather artifacts in Shahrekord Museum, Iran. Abdel-Maksoud [102] identified Penicillium spp., *Aspergillus* spp., and *Fusarium* sp. as the most profuse fungi found on a leather book binding of a Quranic manuscript from the nineteenth century. Nigam et al. [103] attributed degradation of leather and feather objects in Indian museums to keratinophilic and non-keratinophilic fungi.

6. Fungal growth on wool

Common method of isolating dermatophytes of so-called "keratinophilic" fungi is by baiting with hair [104, 105]. Little is known of the relationship of these non-dermatophytic fungi to the decomposition and utilization of keratinous substrates, although previous studies [78] have indicated that role in the breakdown of keratinous tissue may be significant.

Fungal succession on woolen baits was studied and was found that the initial colonizers on woolen baits are non-keratinophilic fungi, while the late colonizers are keratinophilic fungi comprising six phases during fungal succession. The successional tendencies obtained during degradation of wool in samples collected from plain and hilly areas, apart from for the prevailing colonization in the last phase, composed of *chrysosporium tropicum* for the plain, but *Miorosporum* gypseum and *M. fulvum* for the hilly area.

Various workers [68, 69, 75, 76, 106–109] studied degradation of keratinous material, but none of them reported the involvement of other soil microbes in the decomposition of keratin. On succession of fungi on keratinous material [104], little but adequate information on fungal succession on woolen baits is still lacking.

The archeological textiles have a unique position in archaeology, textiles probably contain much archeological information, even more than ceramics. During the Bronze Age (1700–500 BCE), the trade of wool textiles was complex, widespread, significant, and possibly important as the metals [110]. Foundation of economic and political development in the late Middle Ages (c. CE 1100–1500) [111] was Trans-European trade of these materials.

7. Conclusions

The keratinophilic fungi can cause superficial mycosis both in humans and animals. They include a variety of taxonomic groups of filamentous fungi, one of them being the dermatophytes fungi. The keratinophilic fungi can produce a specific enzyme named keratinase that is responsible for keratin degradation. Keratinases can be serine proteases or metalloproteases [86]. The keratinophilic fungi could use keratin from keratinized materials (superficial layers of the skin, hair shaft and nails in humans and claws, horns, wool in animals) as the unique source of carbon and nitrogen. Keratin is found predominantly in feathers, hair, nails, horns, hooves, furs, claws, bird beaks, skin and consists of two types of keratin: α (alpha) and β (beta)-keratin; α keratin (soft) is usually found in hair, wool, horns, nails, claws, and hooves, whereas β keratin (harder) is found in bird feathers, beaks, and claws. Understanding of abiotic and biotic factors and role of different substrates for the growth of these fungi can help researchers to study their growth pattern and conduct their studies for better management of these fungi.

Conflict of interest

The authors declare no conflict of interest.

Author details

Manish Mathur[1] and Neha Mathur[2]*

1 Amity University, Lucknow, Uttar Pradesh, India

2 Amity Institute of Pharmacy, Lucknow, Amity University, Noida, Uttar Pradesh, India

*Address all correspondence to: nmathur1@amity.edu; neha07.mathur@gmail.com

References

[1] Korniłłowicz-Kowalska T, Bohacz J. Biodegradation of keratin waste: Theory and practical aspects. Waste Management. 2011;**31**:1689-1701. DOI: 10.1016/j.wasman.2011.03.024

[2] Kunert J. Physiology of keratinophilic fungi. In: Kushwaha RKS, Guarro J, editors. Biology of Dermatophytes and Other Keratinophilic Fungi. 1st ed. Bilbao: Revista Iberoamericana de Micología; 2000. pp. 77-85

[3] Onifade AA, Al-Sane NA, Al-Musallam AA, Al-Zarban S. A review: Potentials for applications of keratin-degrading microorganisms and their enzymes for nutritional improvement of feathers and other keratins as livestock feed resources. Bioresource Technology. 1998;**66**:1-11. DOI: 10.1016/S0960-8524(98)00033-9

[4] Kim JM, Lim WJ, Suh HJ. Feather-degrading Bacillus species from poultry waste. Process Biochemistry. 2001;**37**: 287-291. DOI: 10.1016/ S0032-9592(01) 00206-0

[5] Vasileva-Tonkova E, Gousterova A, Neshev G. Ecologically safe method for improved feather wastes biodegradation. International Biodeterioration & Biodegradation. 2009;**63**:1008-1012. DOI: 10.1016/j. ibiod.2009.07.003

[6] Farag AM, Hassan MA. Purification, characterization and immobilization of a keratinase from Aspergillus oryzae. Enzyme and Microbial Technology. 2004;**34**:85-93. DOI: 10.1016/j. enzmictec.2003.09.002

[7] Simpanya MF. Dermatophytes: Their taxonomy, ecology and pathogenicity. In: Kushwaha RKS, Guarro J, editors. Biology of Dermatophytes and Other Keratinophilic Fungi. 1st ed. Bilbao: Revista Iberoamericana de Micología; 2000. pp. 1-12

[8] Garg AP, Gandotra S, Mukerji KG, Pugh GJF. Ecology of keratinophilic fungi. Proceedings / Indian Academy of Sciences. 1985;**94**:149-163. DOI: 10.1007/BF03053134

[9] Kushwaha RKS. The genus Chrysosporium, its physiology and biotechnological potential. In: Kushwaha RKS, Guarro J, editors. Biology of Dermatophytes and Other Keratinophilic Fungi. 1st ed. Bilbao: Revista Iberoamericana de Micología; 2000. pp. 66-76

[10] Ni H, Chen Q, Chen F, Ml F, Dong YC, Cai HN. Improved keratinase production for feather degradation by Bacillus licheniformis ZJUEL31410 in submerged cultivation. African Journal of Biotechnology. 2011;**10**:7236-7244

[11] Gröhs Ferrareze PA, Folmer Correa AP, Brandelli A. Purification and characterization of a keratinolytic protease produced by probiotic Bacillus subtilis. Biocatalysis and Agricultural Biotechnology. 2016;7:102-109. DOI: 10.1016/j.bcab.2016.05.00

[12] Lopes FC, Dedavid e Silva LA, Daroit DJ, Velho RV, Pereira JQ, et al. Production of proteolytic enzymes by a keratin degrading *Aspergillus niger*. Enzyme Research. 2011;**2011**:487093. DOI: 10.4061/2011/487093

[13] Syed DG, Lee JC, Li WJ, Kim CJ, Agasar D. Production, characterization and application of keratinase from Streptomyces gulbargensis. Bioresource Technology. 2009;**100**:1868-1871. DOI: 10.1016/j.biortech.2009.08.026

[14] Kannahi M, Ancy RJ. Keratin degradation and enzyme producing ability of Aspergillus flavus and Fusarium solani from soil. Journal of Chemical and Pharmaceutical Research. 2012;**4**:3245-3248. DOI: 10.1139/ w06-067

[15] Sowjanya NC, Chary CM. Degradation of few avean feathers by Microsporum gypseum. Journal of Phytology. 2012;**4**:21-23

[16] Vanbreuseghem R. Technique biologique pour 1' isolement des dermatophytes du sol. Annales de la Societe belge de medecine tropicale. 1952;**32**:173-178

[17] Ajello L. Natural history of the dermatophytes and related fungi. Mycopathologia et Mycologia Applicata. 1974;**53**:93-110

[18] Hejtmanek M, Hejtmankova N, Kunert J. On the occurrence of geophilic dermatophytes in Asia. Ceska Mykol. 1973;**27**:139-161

[19] Canizares O. Dermatology in India. Archives of Dermatology. 1976;**112**: 93-97

[20] Palsson G. Gcophilie dermatophytes in the soil in Sweden. Studies on their occurrence and pathogenic properties. Acta Veterinaria Scandinavica. 1968;**25**:89

[21] Philpot CM. Geographical distribution of dermatophyte: A review. Epidemiology & Infection. 1978;**80**:301-314

[22] Vanbreuseghem R, De Vroey C. Geographic distribution of dermatophytes. International Journal of Dermatology. Apr-Jun 1970;**9**(2):102-109. DOI: 10.1111/j.1365-4362.1970.tb04587.x. PMID: 5426624

[23] Ajello L. The dermatophyte *M. gypseum* as a saprophyte and parasite. The Journal of Investigative Dermatology. 1953;**21**:157-171

[24] Ajello L. Occurrence of Histoplasma capsulatum and other human pathogenic molds in Panamanian soils. The American Journal of Tropical Medicine and Hygiene. 1954;**3**:897-904

[25] Alteras I, Feuerman EI, Grunwald M, Shvili D. Tinea capitis due to Mlicrosporum canis in infants. Mycopathologia. 1984;**86**:89-92

[26] George LK. Animal Ringworm in Public Health. Publication No.727. Washington DC: United States Government Printing Office; 1960. p. 57

[27] English MP, Morris P. *Trichophyton mentagrophytes* var. *erinacei* in hedgehog nests. Sabouraudia. 1969;**7**:118-121

[28] Hubalek Z. Keratinophilic fungi or wild birds. Mykosen. 1972;**15**:207-211

[29] Hubalek Z. Occurrence of keratinophilic fungi in nests of tree sparrow (f'esser montozirs) in relation to the substrate moisture. Ceska Mykol. 1976a;**30**:106-109

[30] Hubalek Z. Interspecific a&nity among keratinolytic fungi associated with birds. Folia Parasital. 1976b;**23**: 267-272

[31] Pugh GJF. Dispersal of *Arihroderma curreyi* by birds, and its role in the soil. Sabowaudia. 1964;**3**:275-278

[32] Pugh GJF. Associations between birds, nests. their pH and keratinophilic fungi. Sabouraudia. 1966b;**5**:49-53

[33] Pugh GJF. Fungi on birds in India. Journal of the Indian Botanical Society. 1966a;**4S**:296-303

[34] Pugh GJF. The contamination of birds of feathers by fungi. Ibis. 1972;**14**:172-177

[35] Sur B, Ghosh GR. Keratinophilic fungi from Orissa, India. I. Isolation from soils. Sabouraudia. 1980;**18**: 269-274

[36] Alteras I, Nesterov V, Ciolofan I. The occurrence of dermatophytes in wild animals from Romania. Sabouraudia. 1966;**4**:215-218

[37] Siniordova M, Hejtmanek M. Orraiatophytes and other keratinopbytic fungi in surface and waste waters. Mykosen. 1970;**13**:467-471

[38] Balbaiioff VA, Usunov P. Organic waste matter of plant origin—Natural source ot' primitive dermatophytes. I. Communication. Mycopathologia et Mycologia Applicata. 1967;**33**:43-48

[39] Caretta G, Frate D. Funghi cheratinofilie dell' isola di Montecristo: isolamenti da sualo. escrementi e pelo di arlimali e piume di uecelli. Chi Psi. Rot. Unit. Lah Criitogam Parlas. 1976;**6**:203-208

[40] Battelli G, Bianchedi M, Frigo W, Amorati P, Mantovani AC, Pagliani A. Survey of keratinophilic fungi in alpine marmat (Marmata marmata) burrow soil and in adjoining soils. Sabouraudia. 1978;**16**:83-86

[41] Pugh GJF, Allsopp D. Microfungi on Signy Island, South Orkney Islands. Br. infarct. Sum. Bull. 1982;**57**:55-67

[42] Buchnicek J. Ober den einfluss der langwelligen UV-strahlung auf da's wachstums von dermatophytes III. Symp. int. Dermetol. Bratislava Abs. 1966:10-16

[43] Berde K. Verhalten von Fadenpilzkolonieo unter der Wirkung des Lichter. Archiv für Dermatologie und Syphilis. 1929;**158**:35-50

[44] Dickey RF. Investigative studies in fungicidal powers of photodynamic action. Journal of Investigative Dermatology. 1961;**39**:7-19

[45] Bohme H, Ziegler H. Verbreitung und Keratinophilie von Anixiopsis siercoraria (Hansen) Hansen. Archiv für Klinische und Experimentelle Dermatologie. 1965;**223**:422-428

[46] Ziegler H. Ektoenzyme der dermatemyzeten. II. Miit. Arch. Klin. Exp. Dermatol. 1966;**226**:282-299

[47] Chmel L, Hasilikova A, Hrasko J, Vlacilikova A. The influence of some ecological factors on keratinophilic fun•pi in soil. Sabouraudia. 1972; **10**:26-34

[48] Kunert J. Utilization of L- and D-cystine by the fungus Microsporum gypseum. Folia Microbiologica. 1982;**27**:390-394

[49] Kunert J. Inorganic sulphur sources for the growth of the dermatophytes Microsporum gypseum. Folia Microbiologica. 1981a;**26**:196-200

[50] Kunert J. Organic sulphur sources for the growth of the dermatophyte Microsporum gypseum. Folia Microbiologica. 1981b;**26**:201-206

[51] Hubalek Z, Rush-Munro FM. A dermatophyte from birds: Microsporum ripariae sp. nov. Sabauraudia. 1973;**11**:287-292

[52] Ajello L, Padhye A. Ktratinophilic fungi of the Galapagos Islands. Mykosen. 1974;**17**:239-243

[53] Das SK, Banerjee AB. Effect of undecanoic acid on phospholipid metabolism in Trichophyton rubnim. Sabouraudia. 1982;**20**:267-272

[54] Hajini GH, Khandari KC, Mohapatra LN, Bhutani LK. Effmt of hair oil and fatty acids on growth of dermatophytes and their in uitro penetration of human scalp half. Sabowaudia. 1970;**8**:174-176

[55] Padhye AA, Mishra SP, Thirumalachar MJ. Occurrence of soil inhabiting dermatophytes and other keratinophilic fungi from soil in Poona. Hindustan Antibiotic Bulletin. 1966;**9**:90-93

[56] Pugh GJF. Cellulolytic and keratinophilic fungi recorded on birds. Sabouraudia. 1965;**4**:85-91

[57] Ajello L, Varsavsky E, Satgin G, Mazzoni A, Mantovani A. Survey of soils for human pathogenic fungi from the Emilia-Romagna region of Italy. Mycopathologia et Mycologia Applicata. 1965;**26**:65-71

[58] AcdaMN. Waste chicken feather as reinforcement in cement- bonded composites. Philippine Journal of Science. 2010;**139**:161-166

[59] Han M, Luo W, Gu Q, Yu X. Isolation and characterization of a keratinolytic protease from a feather-degrading bacterium Pseudomonas aeruginosa C11. African Journal of Microbiology Research. 2012;**6**: 2211-2222

[60] Lin X, Lee CG, Casale ES, Shih JCH. Purification and characterization of a keratinase from a feather-degrading Bacillus licheniformis strain. Applied and Environmental Microbiology. 1992;**58**:3271-3275

[61] Suneetha V, Lakshmi VV. Optimisation of parameters for fermentative production of keratinase by feather degrading microorganisms. World Journal of Microbiology and Biotechnology. 2005;**7**:106-115

[62] Page RM. Observations on keratin digestion by Microsporumgypseum. Mycopathologia. 1950;**42**:591-602

[63] Weary PE, Canby CM, Cowley EP. Keratinolytic activity of Microsporumcanis and Microsporumgypseum. Journal of Investigative Dermatology. 1965;**44**:300-310

[64] Kunert J. Biochemical mechanism of keratin degradation by the actinomycete *Streptomyces fradiae* and the fungus *Microsporum gypseum*: A comparison. Journal of Basic Microbiology. 1989;**29**:597-604

[65] Yu RJ, Harmon SR, Blank F. Isolation and purification of an extracellular keratinase of Trichophyton mentagrophytes. Journal of Bacteriology. Oct 1968;**96**(4):1435-1436

[66] Böckle B, Galunsky B, Müller R. Characterization of a keratinolytic serine proteinase from Streptomyces pactum DSM 40530. Applied and Environmental Microbiology. Oct 1995;**61**(10):3705-3710. DOI: 10.1128/aem.61.10.3705-3710.1995. PMID: 7487006; PMCID: PMC167669

[67] Noval JJ, Nickerson WJ. Decomposition of native keratin by Streptomyces fradiae. Journal of Bacteriology. 1959;**77**:251-263

[68] Kunert J. Keratin decomposition by dermatophytes, I: Sulfate production as a possible way of substrate decom-position. Zeitschrift für Allgemeine Mikrobiologie. 1973a;**13**(6):489-498

[69] Kunert J. Keratin decomposition by dermatophytes. Zeitschrift für Allgemeine Mikrobiologie. 1973b;**13**:189-198

[70] Ruffin P, Andrieu S, Biserte G, Biguet J. Sulfitolysis in keratinolysis: Biochemical proof. Sabouraudia. 1976;**14**:181-184

[71] Stahl WH, MsQue B, Mandels GR, Siu GH. Studies on microbiological degradation of wool, I: Sulfur metabolism. Archives of Biochemistry. 1949;**20**:422-432

[72] Chesters CGE, Mathison GE. The decomposition of wool keratin by Keratinomyces ajelloi. Sabouraudia. 1963;**2**:225-237

[73] Ziegler H, Bohme H. Untersuchungen iiber den Haarabbau durch Dermatomyzeten. Dermat Wschr. 1963;**148**:429-454 Original not seen. Cited by Kunert, Z Allg Microbiol. 1973;**13**(6):489-498

[74] Allsopp D, Seal KI. Introduction to Biodeterioration. London: Edward Arnold; 1986. pp. 1-136

[75] Nigam N, Kushwaha RKS. Biodegradation of keratinous substrates. Biodeierior. of Cultural Propert y, Japan. 1992a:180-185

[76] Nigam N, Kushwaha RKS. Biodegradation of wool by Chrysosporium keratinophilum acting singly or in combination with other fungi. Transactions of the Mycological Society of Japan. 1992b;**33**:481-486

[77] Nigam N, Kushwaha RKS. Occurrence of non-keratinophilic fungi on keratin. Transactions of the Mycological Society Japan. 1989;**30**:l-8

[78] English MP. The saprophytic growth of nonkeratinophilic fungi on keratinized substrate and a comparison with keratinophilic fungi. Transactions of the British Mycological Society. 1965;**48**:219-235

[79] Kishimoto RA, Baker GE. Pathogenic and potentially pathogenic fungi isolated from beech sands and selected soil of Oahu, Hawaii. Mycologia. 1969;**61**:537-548

[80] Carmichael JW. Chrysosporium and some other aleurosporic hyphomycetes. Canadian Journal of Botany. 1962;**40**:l137-l173

[81] Orlita A. Occurrence of fungi on book leather binding from the baroque period. International Biodeterioration Bulletin. 1977;**13**:45-47

[82] Kowalik R. Microbiodeterioration of library materials. Part-2 Micro biodecomposition of basic library materials. Restaurator. 1980;**4**:135-219

[83] Deshmukh SK. Isolation of dermatophytes and other keratinophilic fungi from soil of Mussoorie (India). Mycoses. 1985;**28**:98-101

[84] Garg AK. Isolation of dermatophytes and other keratinophilic fungi from soil of India. Sabouraudia. 1966;**4**:259-264

[85] Jain M, Shukla PK, Shrivastava OP. Keratinophilic fungi and dermatophytes in Lucknow soils and their global distribution. Mykosen. 1985;**28**:148-153

[86] Kushwaha RKS, Agarwal SC. Some keratinophilic fungi and related dermatophytes from soil. Proceedings of the Indian National Science Academy. 1976;**42**:102-110

[87] Nigam N, Kushwaha RKS. Occurrence of keratinophilic fungi with special reference to Chrysosporium species in soils of India. Sydowia. 1990;**42**:200-208

[88] Randhawa HS, Sandhu RS. A survey of soil inhabiting dermato- phytes and keratinophilic fungi of I ndia. Sabourauâia. 1965;**4**:71-79

[89] Gaidau C. Applicative Chemistry of Tanning Metallic Heterocomplexes. Bucharest, Romania: Bentham Science Publishers; 2013

[90] Koochakzaei A, Ahmadi H, Mohammadi Achachluei M. Leather making of Seljuk period at Qohestan of Khorasan (Skin and tanning characterization of leather objects excavated at Ghalee-Kuh-i Ghaen historic site). Quarterly Bulletin of Greater Khorasan. 2012;**3**(7):34-44 [In Farsi]

[91] Ibn al-Athir E. The Complete History. Vol. 6. Tehran: Asaatir; 1995 [In Farsi]

[92] Rajabi N. History and geography of the Ghaen city. Tehran: Shahr-i-Ashoub; 2005 [In Farsi]

[93] Cameron E, Spriggs J, Wills B. The conservation of archaeological leather. In: Kite M, Thomson R, editors. Conservation of Leather and Related Materials. London: Butterworth-Heinemann; 2006. pp. 244-263

[94] Kennedy CJ, Hiller JC, Lammie D, Drakopoulos M, Vest M, Cooper M,

et al. Microfocus x-ray diffraction of historical parchment reveals variations in structural features through parchment cross sections. Nano Letters. 2004;**4**(8):1378-1380

[95] Maxwell CA, Wess TJ, Kennedy CJ. X-ray diffraction study into the effects of liming on the structure of collagen. Biomacromolecules. 2006;7(8): 2321-2326

[96] Creanga DM. Novel aspects in leather covers conservation manuscripts from Punta monastery. European Journal of Science and Theology. 2006;**2**(2):91-97

[97] Lama A, Antunes A, Covington A, Fletcher Y, Guthrie-Strachan J. Red rot in historic leather. Leather International. 2011;**213**(4814):34-38

[98] Thomson R. Testing leathers and related materials. In: Kite M, Thomson R, editors. Conservation of Leather and Related Materials. London: Butterworth-Heinemann; 2006. pp. 58-65

[99] Orlita A. Microbial biodeterioration of leather and its control: A review. International Biodeterioration & Biodegradation. 2004;**53**(3):157-163

[100] Ebrahimi A, Karimi S, Lotfalian S, Majidi F. Allergenic fungi in deteriorating historic objects of Shahrekord Museum, in Iran. Jundishapur Journal of Microbiology. 2011;**4**(4):261-265

[101] Polacheck I, Salkin IF, Schenhav D, Ofer L, Maggen M, Haines JH. Damage to an ancient parchment document by Aspergillus. Mycopathologia. 1989;**106**(2):89-93

[102] Abdel-Maksoud G. Analytical techniques used for the evaluation of a 19th century quranic manuscript conditions. Measurement. 2011;**44**(9):1606-1617

[103] Nigam N, Dhawan S, Nair MV. Deterioration of feather and leather objects of some Indian museums by keratinophilic and non-keratinophilic fungi. International Biodeterioration & Biodegradation. 1994;**33**(2):145-152

[104] Griffin DM. Funal colonization of sterile hair in contact with soil. Transactions of the British Mycological Society. 1960;**43**:583-559

[105] Karling JS. Keratinophilic chytrids. I. Rhizophydrum keratinophilum n. sp. , a saprophyte isolated on human hair and its parasite, Phylctidium mycetophagum n. sp. American Journal of Botany. 1946;**33**:751-757

[106] Kunert J. Keratin decomposition by dermatophytes. Zeitschrift für Allgemeine Mikrobiologie. 1976;**16**:97-105

[107] Kushwaha RKS, Agrawal SC. Microbial degradation of keratin. Proceedings of the National Academy of Sciences, India. 1981;**51**:181-183

[108] Safranek WW, Goos RD. Degradation of wool by saprophytic fungi. Canadian Journal of Microbiology. 1981;**28**:137-140

[109] Wainright MA. A new method for determining the microbial degradation of keratin in soil. Experimentia. 1982;**38**:243-244

[110] Frei KM, Mannering U, Berghe IV, Kristiansen K. Bronze Age wool: Provenance and dye investigations of Danish textiles. Antiquity. 2017;**91**(357):640-654

[111] Von Holstein ICC, Walton Rogers P, Craig OE, Penkman KEH, Newton J, Collins MJ. Provenancing archaeological wool textiles from medieval northern Europe by light stable isotope analysis (δ13C, δ15N, δ2H). PLoS One. 2016;**10**(11):e0162330. DOI: 10.1371/ journal.pone.0162330

Chapter 2

Fungal Growth and Mycotoxins Production: Types, Toxicities, Control Strategies, and Detoxification

Chinaza Godswill Awuchi, Erick Nyakundi Ondari, Ifie Josiah Eseoghene, Hannington Twinomuhwezi, Ikechukwu Otuosorochi Amagwula and Sonia Morya

Abstract

Fungal growth and the production of mycotoxins are influenced by several factors. Environmental conditions such as temperature, water activity, and humidity affect mycotoxin production and fungal growth. Other factors such as pH, fungal strain, and substrate also play roles. Common mycotoxins include aflatoxins, fumonisins, trichothecenes, sterigmatocystin (STC), citrinin, ergot alkaloids, ochratoxins, zearalenones (ZEAs), patulin, deoxynivalenol (DON), Alternaria toxins, tremorgenic mycotoxins, fusarins, cyclochlorotine, sporidesmin, 3-nitropropionic acid, etc. These toxins cause many health conditions in animals and humans, including death. A comprehensive approach starting from the field before planting, continuing throughout the entire food chain is required to control mycotoxin contamination. Good practices, such as proper field practices before and after planting, good harvest practices and postharvest handling, and proper drying and storage measures, help reduce mycotoxin contamination. Several physical, biological, and chemical techniques have been applied to help reduce/eliminate mycotoxin contamination. Food processing also play slight role in mycotoxins removal.

Keywords: Fungal growth, Mycotoxin production, Mycotoxin toxicities, Mycotoxin control and detoxification measures, Factors affecting mycotoxins production

1. Introduction

Fungi are members of the group of eukaryotic organisms that mainly include molds and yeasts. Where conditions are favorable, fungi produce mycotoxins, which are naturally occurring toxic secondary metabolites produced by some fungi, mostly molds, which grow on several crops and foods such as nuts, apples, grains, coffee, fruits, spices, etc., before and after harvest. These filamentous fungi are among the microorganisms that metabolize many organic substances such as sugars, lipids, proteins, etc. Fungi are naturally abundant and ubiquitous and have the capability to attack crops in field, after harvest, and during storage, and can survive under several environmental conditions including humidity, temperature, pH, water activity (a_w), etc. [1, 2]. Fungi commonly invade the commodities consumed by animals and humans, and due to their growth on the commodities, they produce low molecular weight secondary metabolites called mycotoxins [3]. Mycotoxins have been recognized as emerging toxins of concern worldwide [4]. Although more than 100,000 fungal species are known, only few, such as species of *Aspergillus, Fusarium, Penicillium,* etc., are known to produce most of the mycotoxins that significantly affect agriculture, humans, and animals [5].

At present, at least 300 mycotoxins are known, with the widely varied fungal origin, function, structure, toxic potency, and biological effects, although only a few have established significant effects on agriculture, animals, and humans [6, 7]. Many of them have not been sufficiently studied. All mycotoxins identified have between four carbon and complex carbons, which is mainly a result of the different biosynthetic pathways involved in their production [3]. As the production of mycotoxin has not been reported to have any significant biological effect on the growth of fungi, they might play a role in defense mechanisms against several intruders such as insects, animals, nematodes, microorganisms, and even humans [8, 9]. Production of mycotoxins may play role in the maintenance of cell oxidative status at a level essential for the safety of fungi [10]. Several mycotoxins have numerous toxic effects on animals and humans; they pose a real health concern to the public, as they are widely spread in foods worldwide [11].

Mycotoxigenic and pathogenic fungi are common in almost all regions worldwide. They invade, colonize, and grow on many crops, producing mycotoxins under various conditions such as environmental conditions [1]. Several factors have effects on the growth of fungi and production of mycotoxin, and in general, contamination with mycotoxins occurs at various points in the food chain [1]. Fungi presence, however, does not automatically signifies subsequent mycotoxin production, as the conditions required for mycotoxins production are definitive and independent from the conditions that promote the growth of fungi [12, 13]. This chapter throws insight into fungal growth and consequent production of mycotoxins, including their types and toxicities. The chapter also provides methods and strategies for mycotoxins control and detoxification.

2. Fungal growth conditions and mycotoxin formation

Several factors influence fungal invasion, colonization, growth, and consequent production of mycotoxins [14]. The most significant conditions favorable for growth of fungi and production of mycotoxin include temperature and a_w. The optimum temperature for production of mycotoxin by several molds range from 20 to 30°C. Generally, in the tropics and subtropics which are characterized by warm climate, aflatoxins B1, B2, G1, G2, M1, and M2 (AFB1, AFB2, AFG1, AFG2, AFM1, and AFM2 respectively) are of main concern, whereas fusariotoxins, e.g. trichothecenes occur mostly in regions with moderate climate [15, 16]. Additionally, stress factors, including mechanical damage, insect ingression, weed competition, high crop densities, poor fertilization, and drought, can weaken the natural defense mechanism of plants and, as a result, promote fungal colonization, mycotoxin production, and formation of toxins. The optimum conditions for production vary along with temperature, substrate, humidity, and type of mycotoxin [17–19]. Interactions between temperatures and a_w with respect to *F. Verticillioides* growth and mycotoxins production have been studied [17]. The study reported that at 0.995 a_w, the optimal *F. verticillioides* growth rate ranged from 20 to 25°C, however, when the a_w reduced to 0.98, the optimal temperature for growth shifted to 30–35°C. The conclusion was opposite for the production of mycotoxins. For example, the optimum a_w and temperature for the production of fumonisin B1 (FB1) were 0.98–0.995 and 20°C respectively [17]. This shows that the optimum conditions required for mycotoxins production differ from those required for their growth. **Figure 1** shows the minimum, optimum, and maximum required by some fungi for the production of mycotoxins.

The growth of Fungi is categorized into primary and secondary growths. In primary growth, organic compounds are required for the biomass synthesis and production of energy needed to drive chemical reactions to produce the primary metabolites that are essential for growth; secondary growth takes place after the phase of maintained growth and may, sometimes, lead to sporulation and secondary metabolites production [1]. The secondary metabolites, including mycotoxins, have no significant impacts on the fungal growth, but appear to be produced as a result of the excess accumulation of the precursors of primary metabolites, as a means to reduce their concentrations in the fungi [12]. As the fungi that produce mycotoxin and their targeted hosts are diverse in nature, it is difficult to define a single set of conditions which ultimately leads to mycotoxin production. However, in general, the main factors which affect the production of mycotoxins, include temperature, relative humidity, aw, substrate, fungal strain, and pH.

2.1 Temperature, relative humidity, and water activity

Environmental factors play significant roles in determining the occurrence of fungi; the fungal activities and colonization are predominantly determined by conditions such as temperature and humidity [21]. These factors influence the mycotoxigenic fungal prevalence, development, frequency, distribution, and survival, and their subsequent accumulation of toxin. Also, humidity and temperature affect plant growth, health, strength, and influence the mycotoxigenic fungal competitiveness. Due to differences in growth requirements and the environmental factors, fungal development and production of mycotoxin differ from one geographical region to another [22–24]. Before harvest, on the field, fungi *Fusarium* species mostly dominate since they are hygrophilic and require at least 90% relative

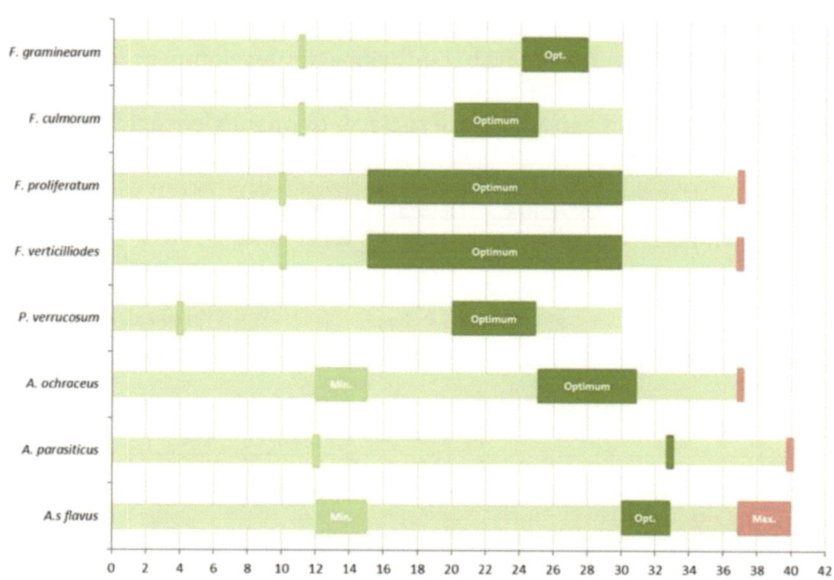

Figure 1.
Minimum, optimum, and maximum range of temperatures (°C) for the production of mycotoxins (adapted from [20]).

humidity (RH) to germinate/grow. Whereas after harvest, the hygrophilic fungi are not seen, as xerophylic and mesophilic fungi, such as *Penicillium spp.* and *Aspergillus spp.*, germinate and grow, leading to mycotoxins production at 80% or less RH and 80 to 90% RH, respectively [23]. In storage, if the surrounding environment's RH is more than the food's equilibrium RH, the food gains moisture and its a_w increases [25]. Increased a_w during storage increases the food vulnerability to fungal invasion, germination, growth, and production of mycotoxin. The major optimal factors for fungal growth and mycotoxin production are shown in **Table 1**.

For temperature requirement, most fungi are mesophilic and grow in range of temperature between 5 and 35°C, with optimal growth occurring around 25–30°C [1]. In addition, there are species of fungi called psychrophiles (tolerant to low temperatures) and thermophiles (can bare high temperatures). When there is shift in optimal temperature range (see **Table 1**), it may lead to a halt in growth [28]. The conditions that encourage the growth of fungi may not necessarily result in the production of mycotoxins. However, temperatures of 25–30°C, a_w above 0.78, and RH of 88–95% favor fungal growth and production of mycotoxin [23, 25]. The most common mycotoxins, their toxicities, and fungi that produce them are shown in **Table 2**. Most of these mycotoxins are carcinogenic, genotoxic, mutagenic, immunotoxic, teratogenic, etc.

2.2 Fungal strains

There is variation in the toxicity of Fungal species and their mycotoxins production may usually depends on species, strains, and/or genera. The production of mycotoxin is influenced by strain stability, variation, and specificity [29]. Strains of the same species can have different optimum conditions required for growth and production of mycotoxins; also, strains of the same species can produce one or more different mycotoxins. For instance, while *Aspergillus flavus* can thrive within 15 to 44°C and produce aflatoxin B1 (AFB1), *Aspergillus carbonarius* thrives at 8 to 40°C and produce ochratoxin A (OTA) [23].

2.3 pH

The surrounding medium and its pH influence the development of fungi and production of mycotoxins. pH of the surrounding medium affects fungal growth either by direct or indirect actions on cell surfaces or on nutrient availability

Fungi	Water activity	Temperature for growth	Optimal water activity	Optimal temperature	Optimal water activity for mycotoxin production	Optimal temperature for mycotoxin production
Aspergillus carbonarius	0.90–0.93	8–40°C	0.94–0.99	32–35°C	0.98	30–35°C
Aspergillus flavus,	0.91–0.99	15–44°C	0.95	35°C	0.99	33°C
Aspergillus ochraceous	0.80–0.98	10–40°C	0.96–0.98	24–31°C	0.98	25–30°C
Aspergillus parasiticus	0.91–0.99	15–44°C	0.95	35°C	0.99	33°C

Table 1.
Optimal conditions for fungal growth and mycotoxin production [26, 27].

Mycotoxins	Common fungal species	Foods commonly found	Toxicity
Aflatoxins (Aflatoxins B1, B2, G1, G2, M1, M2)	*Aspergillus flavus*, *Aspergillus parasiticus*, *Aspergillus bombycis*, *A. pseudotamarii*, *A. nomius*, etc.	Cereals, seeds, vegetables, nuts, legumes, fruits, etc.	Liver cancer; target DNA; hepatocellular carcinoma; mutagenic and teratogenic effects. Aflatoxin M1 is a metabolite of aflatoxin B1 and is commonly found in milk and dairy products
Alternaria toxins (altertoxin, alternariol methyl ether, alternariol, altenuene, tentoxin, tenuazonic acid)	*Alternaria* species such as *Alternaria solani*, *Alternaria japonica*, *Alternaria dauci*, *Alternaria triticina*, *Alternaria tenuissima*, *Alternaria brassicae*, *Alternaria alternata*	Fruit, grains, beer, fruit juices, vegetables, seeds, vegetable juices, wine, peppers, tomatoes, dried fruit, flour, bran, wheat, cereal products (e.g. rice and oat flake), sunflower seeds, sunflower oil, etc.	Although most Alternaria toxins show low acute toxicities, alternariol methyl ether and alternariol are genotoxic, cytotoxic, carcinogenic, and mutagenic effects, with scientific-based findings from toxicological studies, in vitro, involving mammalian and bacterial cells. Tenuazonic acid has phytotoxic and antibacterial properties and acute toxicities for dogs, chicken, and mice; it also causes hematological disorders in humans.
Emerging fusarium mycotoxins (enniatins, NX-2 toxin, beauvericin, moniliformin, fusaproliferin, etc.)	*Fusarium* species, such as *F. verticillioides*, *F. subglutinans*, *F. proliferatum*, *F. arthrosporiodes*, *F. chlamydosporum*, *F. redolens*, *F. acuminatum*, *F. avenaceum*, *F. oxysporum*, *F. beomiforme*, etc.; *Beauveria bassiana*	Corn, rice, corn products, seeds, nuts, coffee, tree nuts, dried fruits, beans, vegetable oil, etc.	They are potentially toxic to humans and animals. Beauvericin has antifungal, insecticidal, and antibacterial properties, and may have toxic effects leading to apoptosis induction, increased cytoplasmic calcium concentration, and fragmentation of DNA in cell lines of mammals.
Ergot alkaloids	Clavicipitaceae (e.g. *Neotyphodium* and *Claviceps*) and Trichocomaceae (e.g. *Penicillium* and *Aspergillus*) families. *Claviceps purpurea* is the dominant producer	Rye, barley, wheat, triticale, oats, etc.	Ergot alkaloids are both harmful and beneficial to humans. Causes ergotism. Gangrenous and convulsive forms of toxicities. Can cause delirious seizures, St. Anthony's Fire, and fits. Can cause
Fumonisins (fumonisins B1, B2, B3, etc.)	species of *Fusarium* (such as *F. verticillioides*, *F. nyagamai*, *F. fujikuroi*, *F. Proliferatum*, *F. oxysporum*, *F. Globosum*, etc.), *Aspergillus awamori*, *A. niger etc.*	Corn, corn products, asparagus, rice, beer, soybeans, beans, sorghum, etc.	Inhibits sphingolipids synthesis. Linked to esophageal and liver cancer in human, atherosclerosis in monkeys, equine leukoencephalomalacia in horse, porcine pulmonary edema and pulmonary artery hypertrophy in swine, and kidney and liver cancer in rodents.
Ochratoxins (Ochratoxins A, B, C)	*Aspergillus* and *Penicillium* genera, such as *Aspergillus ochraceus*, *Aspergillus carbonarius*, *Aspergillus niger*, *Penicillium verrucosum*, etc.	Cereals, seeds, fruits, legumes, vegetables, nuts, etc.	Ochratoxins have immunotoxic, neurotoxic, hepatotoxic, teratogenic, and nephrotoxic activities; cause nephropathy in pigs; In human, ochratoxin A was linked to urothelial tumor, renal failures, chronic interstitial nephropathy, and Balkan endemic nephropathy.

Mycotoxins	Common fungal species	Foods commonly found	Toxicity
Patulin	*Penicillium expansum*, *A. clavatus*, *Penicillium patulum*, *Penicillium griseofulvum*, *Penicillium urticae*, *Penicillium crustosum*, etc.	Apples, apple products, fruits, cereals, legumes, vegetables, nuts, seeds, etc.	Mutagenicity, carcinogenesis, immunotoxicity, teratogenicity, and neurotoxicity are chronic and acute effects of patulin showed on cell cultures. Patulin has neurotoxic and immunotoxic effects in animals, but no reliable evidence has shown its carcinogenicity to human, although studies have shown human toxicities, such as hemorrhage, ulceration, vomiting, and nausea
Sterigmatocystin	Species of *Aspergillus*, such as *A. versicolor* (major producer), *A. aureolatus*, *A. amstelodami*, *A. ruber*, *A. Chevalieri*, *A. sydowi*, *A. quadrilineatus, etc.* Also produced by species of *Emiricella*, *Chaetomium*, *Penicillium,* and *Bipolaris*	Peanuts, corn, wheat, grain products, barley, rice, etc.	Sterigmatocystin has teratogenic, mutagenic, and carcinogenic effects. Can cause hepatic toxicity in most animals; bloody diarrhea and death in cattle; hepatocellular carcinoma and squamous cell carcinomas in rats; LD50 in mice is ≥800 mg/kg
Trichothecen es, e.g. deoxynivalen ol (vomitoxin), anguidine, T-2 toxin, 3- and 15- acetyldeoxyni valenol, nivalenol, HT-2 toxin, crotocin, diacetoxyscir penol, macrocyclics, etc.	Species of *Fusarium* (such as *Fusarium crookwellense, F graminearum, F poae, F culmorum*), *Myrothecium*, *Trichoderma*, *Cephalosporium*, *Stachybotrys*, *Spicellum*, *Verticimonosporium*, *Trichothecium*	Rice, oats, wheat, vegetables, rye, barley, maize, etc., and animal foods, such as liver, eggs, milk, and kidney	Trichothecenes can diffuse into cells, and block translation by interacting with eukaryotic ribosomes. Trichothecenes exposure affect nearly all key systems in vertebrates, cause alimentary toxic aleukia (ATA) in humans, etc. They inhibit DNA, RNA, and protein synthesis, and also cause lipid peroxidation, apoptosis, inhibit mitochondrial functions, cause changes in neurotransmitters, and cytokine activation.
Zearalenone (formerly referred to as F-2 toxin)	*Fusarium* species, such as *F. crookwellense*, *F. cerealis*, *F. semitectum*, *F. equiseti*, *F. graminearum*, *F. culmorum, etc.*	Maize, soybean, oats, barley, wheat, rice, rye, sorghum, grain products, etc.	Zearalenone chronic administration can cause uterine fibroids, pituitary adenomas, hepatocellular carcinoma, and liver damage in mice, and chronic progressive hematotoxicity, testicular atrophy, cataracts, retinopathy, and nephropathy in rats; Among other animals, pig is more prone its toxicities. Zearalenone or its metabolic compounds are known to bind transcription factors, including pregnane X receptor involved in expressing enzymes in pathways of biosynthesis

DOI: http://dx.doi.org/10.5772/intechopen.100207

Mycotoxins	Common fungal species	Foods commonly found	Toxicity
Other common mycotoxins (Fusarins [fusarins A– F], Tremorgenic mycotoxins, Cyclochlorotine, Sporidesmin, 3-nitropropionic acid,)	Fusarins are produced by the species of *Fusarium*, such as *Fusarium verticillioides* (formerly *Fusarium moniliforme*), *Fusarium graminearum* (*Fusarium venenatum*), *Fusarium poae*, *Fusarium sporotrichioides*, *Fusarium oxysporum.* Tremorgenic mycotoxins are produced by *Aspergillus terreus,* species of *Penicillium* genus, etc.; *Pithomyces chartarum* produces Sporidesmin; Cyclochlorotine is produced by *Penicillium islandicum*; 3-nitropropionic acid (3-NPA) is produced by the species of *Arthrinium*;	Many foods and feeds	Fusarins are mutagenic; 3-nitropropionic acid interjects mitochondrial electron transport; Cyclochlorotine interrupts myofibrils and is hepatotoxic in animals; Due to the hydrophobicity of sporidesmin, it can be integrated easily into the membranes of cells, in which it changes the organization of the bilayer Tremorgenic mycotoxins cause “staggers syndromes” in livestock, and are linked to neurological conditions, such as seizures, tremors, mental confusion, and even death in humans.

Table 2.
Common mycotoxins, their producing fungi, and known toxicities [1, 2, 25].

respectively. Fungi can modulate the pH of the surrounding medium through secreting alkali or acids; species of *Aspergillus* and *Penicillium* can acidify the surrounding through citric and gluconic acids secretion [30]. The ability to control the pH provides fungi a better possibility of surviving in their host. In addition, the pH can have influence on the interactions of temperature and a_w, as it affects metabolic processes, including morphogenesis and sporulation [31].

The pH can also affect the gene expression for biosynthesis, e.g. at pH 8, the genes responsible for production of ochratoxin A by *Penicillium verrucosum* are expressed [32]. Although the pH effect on the production of some mycotoxins has not been fully established for every type of mycotoxins, however acidic conditions are known to promote germination and production of mycotoxin. Production of aflatoxins requires pH 4.0 and the pH has inverse relationship with the level of synthesis [12]. In the same way, OTA levels are higher when *Aspergillus ochraceus* are at low pH [32]. Fumonisin B1 (FB1) is unstable in alkaline medium and requires pH 4.0–5.0 for synthesis; production of trichothecenes is initiated in acidic conditions [12].

2.4 Substrate

Mycotoxigenic fungi grow on several substrates, however, the major reason for their predominate on some foods has not been sufficiently established. The nutrients needed for the fungal growth, mostly nitrogen and carbon, are commonly found in foods, especially those rich in carbohydrates, molds are found in many foods [33]. Substrates that encourage the growth of fungi may not necessarily be support mycotoxin production, as conditions that promote the production of mycotoxins are usually different from those needed for fungal growth (see **Table 1** and **Figure 1**). Generally, mycotoxins production is greatly influenced by the interactions between many factors in substrate, such as temperature, a_w, pH, and composition (e.g. simple sugars). The substrate's osmotic pressure affects the growth of fungi and the production of mycotoxin, and several studies reported that it can aid in evaluating fungal physiological responses, as well as can affect the secondary metabolites biosynthesis such as mycotoxins biosynthesis [34]. On the other hand, upon osmotic stress fungal species adjust their physiological responses to enhance their survival and adaptation [34].

Sugars have carbon and filamentous fungi have the natural ability to hydrolyse several sources of carbon to support growth and produce energy [35]. Consequently, in sugars presence, such as the presence of simple sugars, which readily breakdown, there is higher frequency of fungal growth. When complex sugars dominate, fungal growth is slower as the complex sugars need further digestion to yield simpler units of carbon that are readily absorbable. Simple sugars may contribute to the mycotoxins production. [36] reported that increase in the concentration of soluble sugars to 3 and 6 percent, especially maltose, glucose, and sucrose, promoted production of AFB1 in cell cultures. [37] also reported that more production of AFB1 by *Aspergillus flavus* resulted from an increase in the medium sugar levels.

2.5 Effects of climate change on mycotoxins production

Climate change has resulted in changes in most environmental conditions, including rise in global temperature which is expected to increase by 1.5–4.5°C by 2100 [38]. A rise in droughts, precipitation, flooding, and extreme weather conditions are expected [39]. Climate change and global warming affect food security greatly, including reduction in crop quality, reduction in yields, and increased food safety challenges making some crops unsafe for human and animal consumption. Global change affects mycotoxin production, mainly by affecting the environmental conditions that influence their production [40].

Climate change affects different regions in various ways, with some regions having advantage whereas the opposite is the case for other regions [41]. The Mediterranean basin and Southern Europe most likely experience significant changes resulting in increase in mycotoxin prevalence, while the effects of climate change are anticipated to be positive in northern Europe [42]. As fungal growth, their germination, and production of mycotoxins are largely influenced by environmental conditions, especially the optimal conditions, temperature changes and change in humidity induced by climate change may several effects on production of mycotoxins. The mycotoxins usually produced at low temperature may not be produced at higher levels, whereas others predominant in tropical and sub-tropical regions, including aflatoxins, may be produced in temperate regions as a result of the expected rise in temperatures in these regions; for example, in Italy, in 2003 and 2004, hot and dry conditions resulted in the *Aspergillus flavus* colonization and aflatoxins production [43]. Different mycotoxins can be affected differently, usually based on the optimum conditions required for their production. Climate change also affects mycotoxins production indirectly via the increase of pest and insect populations, global spread, and attacks, early maturing and ripening of crops, decreased plant resilience, and change in host pathology upon the presence of CO2 in the atmosphere [42, 44].

3. Mycotoxins prevention and control

The production of mycotoxins has shown unavoidability and, as a result, many foods are being contaminated regularly. Mycotoxins greatly and widely vary and are produced by many fungi at various stages, on many crops, and consequently, a specific strategy for controlling all the mycotoxins has proven difficult with little or no success in decontaminating the affected foods or reducing all the mycotoxins to safe levels; most specific control strategies may only be effective in reducing the levels of specific types of mycotoxins. However, certain control measures can be employed to prevent or minimize their entrance, production, and occurrence in foods. Currently, no method has proven sufficient to totally control all mycotoxins. A successful strategy may adopt a combination of food safety system involving suitable quality measures at every production stage to reduce the frequency of mycotoxins occurrence in the final food products, which would include taking appropriate measures before, during, and after harvest. **Table 3** shows the overview of the action mechanisms of mycotoxins.

3.1 Mycotoxins control using appropriate field practices

Most fungi are phytopathogens that infect the crops in field, and their pre-harvest management is very important. In general, fungi that mostly predominate in field include *Alternaria spp., Cladosporium spp.,* and *Fusarium spp.* However, *Penicillium spp.* and *Aspergillus spp.* also occur in field at low levels and the contamination levels in general are usually higher anywhere climate conditions are favorable to the production of mycotoxin [45]. Whereas it is unlikely to totally prevent the production of mycotoxins in the field before harvest, it is of extreme importance to adhere to strategies which aim to reduce contamination to the barest possible minimum in preharvest. To choose and implement suitable strategies, a sufficient knowledge about the mycotoxigenic fungi, crops mostly affected, harvesting

Action	Mechanism
DNA effects	There are two major types of interactions between nucleic acids and mycotoxins; reversible and noncovalent or irreversible and covalent. The covalent and irreversible interaction between DNA and AFB1 results in the formation of N7-guanine adduct.
Effects on hormones	ZEA has structural similarity to 17β-estradiol; the effects of ZEA on receptors of estrogen explain its fertility problems on humans and animals. Ergovaline, an ergot alkaloid, reduces levels of prolactin in animals by acting as an agonist of dopamine.
Epigenetic properties	Few mycotoxins change the levels of DNA methylation
Important metabolic enzymes inhibition	OTA, citroviridin, and AFB1 affect the metabolism of carbohydrates, while rubratoxin B and trichothecenes interfere with metabolism of lipid. Fumonisin B1 inhibits argininosuccinate synthetase. Fumonisins chemical structure has high similarity to those of sphinganine and sphingosine, the sphingolipids backbones. Consequently, fumonisins inhibit ceramide synthase competitively.
Ionophore	Beauvericin and enniatins that are produced by the species of *Fusarium* have ionophoric activities specific to potassium and cause influx of potassium into the matrix of mitochondria, followed by swelling of the mitochondria.
Mitochondrial interactions	By binding covalently to the enzyme active site, 33-NPA permanently inactivates succinate dehydrogenase. Acrebol, from *Acremonium exuviarum*, inhibits mitochondrial complex III, consequently causing ATP depletion by inhibiting the chain of respiration. Fumonisin B1 was found to obstruct the mitochondrial complex I in human neuroblastoma cells and rat primary astrocytes, resulting in reduced cellular and mitochondrial respiration and an increase in reactive oxygen species (ROS) generation, with calcium signaling deregulation.
Necrosis and apoptosis	AFB1 cytotoxic effects in lymphocytes of humans involve necrosis, caspase activation, and apoptosis.
Protein interaction	The plasma albumin binds to aflatoxins. After oxidation of AFB1 by cytochrome P450s, two epoxides are formed and they react with the lysine ε-amino group, forming AFB1–albumin adducts. Aflatoxins are immunosuppressive, and in several studies, they suppress immune response mediated by the cell and impairs phagocytosis and chemotaxis. Most immunotoxic properties of fumonisin B1 may be as a result of its capability of altering the levels of mRNA and/or expression of IL-1β, IFN-γ, and TNF-α in several scientific experiments. Penitrem obstructs uptake of glutamate and GABA (γ-aminobutyric acid) into cerebellar synaptosomes, modulating the function of GABA receptor. One of the ways patulin exerts its toxicities is by causing a dose- and time-dependent phosphorylation increase of c-Jun N-terminal kinase, protein kinases 1 and 2 regulated by extracellular signal, and p38 kinase, contributing to downstream effects including cell death and DNA damage. A mycotoxin known as Secalonic acid D, which causes "cleft palate", phosphorylates the binding protein of cAMP response element.
Ribosomal binding	Ochratoxin A competes with phenylalanine-tRNA ligase and inhibits synthesis of protein; both aspartame and phenylalanine reduce toxicity of OTA by competing with it. Trichothecenes toxicities are due to their capability to bind the eukaryotic ribosomes' 60S subunit and inhibit the reaction of peptidyl transferase.
RNA polymerase effects	AFB1 has inhibitory effects on chromatin-bound RNA polymerase which is DNA-dependent and, consequently interferes with synthesis of RNA. Luteoskyrin and patulin also inhibit RNA polymerase.

Table 3.
Mycotoxins action mechanisms [2].

practices, and proper field management play important role [46]. Many factors such as delayed harvesting, poor soil fertility, heat, insect infestation, and drought contribute to the production of mycotoxins in the field [46, 47]. Appropriate practices in the field include the management and preparation of field before planting, and proper management of crop and field after planting.

3.1.1 Preparation and management of the field before planting

Preparation of the field prior to planting is critical in controlling fungal invasion and the consequent production of mycotoxins. Deep plowing, tilling, production cycle, use of disease-resistant cultivars, crop rotation, use of high-quality seeds, etc. play important role. Deep plowing and tilling can be essentially used for the removal of remaining plant materials. Previous residues of crops which persist on the soil end up deteriorating and harboring soil-borne fungi, which increase their possibility of invading new crops. Plowing puts debris under the ground, making them not accessible to inhabitation by fungi. Tilling may also increase water availability to crops by minimizing the compressed layers of soil [48]. Additionally, rotation of crops prevents the build-up of fungi; it was reported that the production of mycotoxins is higher in lands where same crops are consecutively grown for years, as molds that may colonize a plant can occur from year to year if same crop was continuously planted [46–48]. Seeds for planting are also very important. Seeds with good quality contribute to the health of plants' growth to withstand fungal invasion.

3.1.2 Management of crop and field after planting

Facilitating the healthy plants growth after planting through implementing proper practices in the field and decreasing the stress on crops reduces fungal growth and production of mycotoxins [47, 49]. Fertilizers application improves the health of plants and maintains their disease and fungal resistance. Availability of nutrients is important for plant life and lack of proper nutrition of plant results in a break in the plant stem, exposing it to more invasion by fungi and other microorganisms [50]. Proper irrigation also has the capacity to prevent accumulation of mycotoxins via method and timing of irrigation. Proper timing can prevent drought stress, while the method of irrigation that control splashing can help prevent the spreading of fungi. The control of insect and weed is also important in preventing crop diseases and invasion by fungi [2, 51]. Fungicide application at proper doses can also help in controlling the fungal invasion and consequent production of mycotoxins.

3.1.2.1 Early fungal detection as a control means

Fungi presence does not necessarily indicate mycotoxin production; however, their presence implies an increased mycotoxin production risk if the conditions are suitable for the production of mycotoxin. Consequently, early detection of fungi which allows corrective measures can be very critical in controlling mycotoxin production [52]. When fungi are detected at early stages, the methods of decontamination (see [2]) in the field can be used to prevent fungal germination and growth, which in turn prevent subsequent production of mycotoxins.

3.1.2.2 Biological control after planting

After planting of crops, biological control measure largely includes the applying harmless species of fungi that compete with mycotoxigenic fungi, inhibiting their pathogenic activities. While this measure seems practically challenging, it presents safer control methods that are ecofriendly. This measure implies introducing a strain of harmless bio-agents, e.g. yeasts or bacteria, which compete with mycotoxigenic fungi for resources, thereby reducing their growth and ability to produce mycotoxin. For instance, *Aspergillus spp.* strains that do not produce aflatoxins are applied as biological agents to compete with the strains that produce aflatoxins and prevent their dominance and subsequent aflatoxins production [53, 54]. A study reported that after applying non-aflatoxigenic strain of *Aspergillus parasiticus* on the soil in the field, significant reduction in the levels of aflatoxin was attained [55]. Same was the case after non-toxic strains of *Aspergillus flavus* were applied to cotton row. However, this biological control method has limitations that may discourage its wide application. First of all, the biological agent used may have impact on other natural occurring microorganisms. Secondly, non-toxigenic strains of fungi, even though they can help in reducing the production of mycotoxins, may produce other metabolic compounds that might be toxic to humans and animals. Thirdly, the non-toxigenic strains may result in underestimating the levels of mycotoxins, as they may have effects on the fungal metabolic pathways and cause the production of modified derivatives of mycotoxins [54]. Additionally, the ability to produce mycotoxins may be transferred from one fungi to another (a descendant) via non-toxigenic strains crossing with toxigenic strains, resulting in likely reproduction of successive fungi that produce mycotoxins [54].

3.1.2.3 Chemical control after planting

Use of fungicides as chemical control is one of the current most effective ways to proper control of crop invasion by fungi and consequent production of mycotoxins [53, 56]. Use of chemicals such as captan, mancozeb, cinnamaldehyde, citronella oil, tea tree oil, monocerin, sulfur, etc. has been recommended.

3.2 Appropriate measures during harvest

Another critical stage for controlling mycotoxin production is during harvest, where moisture plays the most crucial role for the protection of crops. Harvest should start after dry weather condition. Harvesting crops in wet weather may make them more vulnerable to the growth of fungi and subsequent mycotoxin production; also, keeping crops on the field for a long period of time can increase the risk of invasion by fungi, birds, rodents, pests, and insects, all of which can contribute to mycotoxin production through one way or the other [1]. It is important to prevent mechanical damage during harvest.

3.3 Postharvest control and prevention

3.3.1 Appropriate storage

Measures should be taken to limit fungal invasion and the production of mycotoxins before food products reach storage facility. If commodities are highly contaminated before reaching storage facility, it is difficult and more complicating to prevent further fungal or mycotoxins accumulation [47]. During storage, proper techniques should be applied to avoid fungal invasion and germination. Several fungi inhabit stored grains, especially fungal species that are infrequent in the field such as *Penicillium spp.* and *Aspergillus spp.*, whereas fungi in the field that require high water activity pose less significance during storage [57]. Several factors and storage conditions affect fungal growth and germination capacity during storage, especially temperature and a_w. The fungi active in storage can grow at 70–90% relative humidity, and they usually thrive at temperatures of 10 to 40°C with 25–35°C range of optimum temperature [58]. Use of chemical preservatives, hygienic conditions, and storage time can also affect fungal growth and mycotoxin production. Temperature and water activity should be monitored and properly controlled during storage. The relative humidity should be kept below 70% throughout storage. The foods should be kept at low temperatures to reduce fungal activity during storage. The temperature of the stored food can be used as a good storage quality indicator.

3.3.2 Chemical detoxification and decontamination

Chemical methods make use of chemical treatments with oxidizing agents, reducing agents, alkalis, and acids which are synthetic or organic. The chemicals are employed in the detoxification of mycotoxins after adding the foods. Chemicals can be introduced through packing, fumigation, mixing, or immersion [59]. Chemical commonly used include formaldehyde, ozone, chlorinating agents, sodium bisulphite, hydrogen peroxide, hydrochloric acid, ammonium hydroxide, organic acids, and natural substances (e.g., spices, herbs, and their extracts) [59–61]. Treatment with chemical is effective in removing some mycotoxins, although they are mostly weak chemicals and most mycotoxins can be resistant to the chemicals. Ozone treatment is promising as it has the ability to degrade mycotoxins via reacting with bonds in the structure of mycotoxins, such as the double bonds in aflatoxin B1. By-products may form in the process [62, 63].

3.3.3 Physical detoxification and decontamination

Physical measures to control mycotoxins mostly comprise separating contaminated and damaged crops from wholesome crops using methods such as sorting, steeping, dehulling, washing, density segregation, sieve cleaning, etc., to reduce the mycotoxins levels [60, 64]. In physical process of decontamination, mycotoxins that are soluble in water can be removed partially from the grain outer surface using water or solutions of water [62]. Physical methods also include mycotoxins destruction and removal using irradiation and heat treatment [59]. Thermal processes including extrusion heating, radiofrequency, microwave, infrared, steam, boiling, etc. have been applied as innovative methods of mycotoxin decontamination [63]. Heat treatments that make use of a combination of the temperature conditions and time might give the most significant approach to control mycotoxins, however, most mycotoxins have heat stability and require very high temperatures and prolonged durations of processing for destruction, which may not be achieved in conventional food processing [64, 65] or destructive to the food constituents such as nutrients [63]. Non-thermal treatments like irradiation might be effective by partial lowering of mycotoxin levels, as the mycotoxins absorb energy of radiation and can be widely applied at industrial scale [62]. However, Irradiation is not widely applied due largely to public distrust of irradiated foods, since radiation has the ability to penetrate cells, causing DNA damage that results in mutations. In spite of that, the European Commission has approved 10 kGy dose as the maximum dose allowable for food application after it was demonstrated that this level poses no danger to humans [62].

Cold plasma is a recent novel non-thermal physical method used for the removal of fungi and mycotoxins. Cold plasma is ionized gas with partially ionized molecules and atoms with net charge approximately zero [66]. Cold plasma treatment caused the destruction of fungal DNA and cell wall, leading to the leakage of intracellular components [66–69]. Cold plasma partially or completely destroyed mycotoxins quickly [66, 70]. The efficiency of mycotoxins destruction mechanism is associated with their molecular structures, the nature of plasma, and subsequent interactions; destruction may be associated with the production of free radicals in the course of the treatment regimen, or the UV photons and ozone presence, or reactive electrons and ions [66]. Treatment with cold plasma is distinctive from conventional methods, as the mycotoxins can be rapidly decontaminated at ambient pressure and temperature conditions without affecting the quality of the food [66]. However, some studies indicated that treatment with cold plasma may have effects on lipids, and it may be difficult to apply at large-scale industries, especially for the treatment of foods with irregular shape and bulky foods [71].

Photocatalytic detoxification is another emerging non-thermal technique used for mycotoxins removal from foods. The process involves chemical reaction initiated by photons absorption by solid photocatalyst, resulting in redox reactions on the photocatalytic material surface, leading to free radicals' formation which interact with the contaminants (mycotoxins) and helps degrade or convert them to lesser toxic substances via oxidation [72]. Several studies have reported photocatalytic detoxification effects on mycotoxins [72–75].

3.3.4 Biological detoxification and decontamination

Biological decontamination strategy makes use of microorganisms (algae, molds, yeasts, and bacteria). The microorganisms may bind, modify, or degrade the mycotoxins to lesser toxic compounds in some feed and foods through decarboxylation, hydrolysis, deamination, glucosylation, and/or acetylation [60]. Ochratoxin A can be converted to phenylalanine by some bacteria, plants, mold, and yeasts [76]. Some enzymes and microorganisms can be added to feed for mycotoxins degradation/detoxification in the ruminants' GI tracts. Yeasts and Lactic acid bacteria are mostly used for decontaminating mycotoxins as they can reduce their levels by binding to their cell surface or by converting them to lesser toxic substances [62]. Additionally, enzymatic catalysis can be used as they have promising applications in decontaminating mycotoxins [77].

4. Conclusion

Fungi commonly invade the commodities consumed by animals and humans, and due to their growth on the commodities, they produce low molecular weight secondary metabolites called mycotoxins. Environmental conditions such as temperature, water activity, and humidity affect mycotoxin production and fungal growth. Other factors such as pH, fungal strain, and substrate also play roles. The conditions that encourage the growth of fungi may not necessarily result in the production of mycotoxins. Common mycotoxins include aflatoxins, zearalenones (ZEAs), patulin, deoxynivalenol (DON), fumonisins, trichothecenes, sterigmatocystin (STC), citrinin, ergot alkaloids, ochratoxins, Alternaria toxins, tremorgenic mycotoxins, fusarins, cyclochlorotine, sporidesmin, 3-nitropropionic acid, etc. These toxins cause many health conditions in animals and humans, including death. A comprehensive approach starting from the field before planting, continuing throughout the entire food chain is required to control mycotoxin contamination. Good practices, such as proper field practices before and after planting, good harvest practices and postharvest handling, and proper drying and storage measures, help reduce mycotoxin contamination. Several physical, biological, and chemical decontamination methods can be used to reduce/eliminate mycotoxin levels. More studies are required to develop methods and techniques that can effectively reduce all mycotoxins in foods and feeds.

Conflict of interest

The authors declare no conflict of interest.

Author details

Chinaza Godswill Awuchi[1,2*], Erick Nyakundi Ondari[1], Ifie Josiah Eseoghene[2], Hannington Twinomuhwezi[2,3], Ikechukwu Otuosorochi Amagwula[4] and Sonia Morya[5]

1 Department of Biochemistry, Kampala International University, Bushenyi, Uganda

2 School of Natural and Applied Sciences, Kampala International University, Kampala, Uganda

3 Department of Chemistry, Kyambogo University, Kampala, Uganda

4 Department of Food Science and Technology, Federal University of Technology Owerri, Owerri, Imo State, Nigeria

5 Department of Food Technology and Nutrition, Lovely Professional University, Phagwara, India

*Address all correspondence to: awuchichinaza@gmail.com; awuchi.chinaza@kiu.ac.ug

References

[1] Daou R, Joubrane K, Maroun RG, Khabbaz LR, Ismail A, El Khoury A (2021). Mycotoxins: Factors influencing production and control strategies. AIMS Agriculture and Food, 6(1): 416-447. DOI: 10.3934/agrfood.2021025

[2] Awuchi, C.G.; Ondari, E.N.; Ogbonna, C.U.; Upadhyay, A.K.; Baran, K.; Okpala, C.O.R.; Korzeniowska, M.; Guiné, R.P.F. (2021). Mycotoxins Affecting Animals, Foods, Humans and Plants: Types, Occurrence, Toxicities, Action Mechanisms, Prevention and Detoxification Strategies—A Revisit. Foods, *10*, 1279. https://doi.org/10.3390/foods10061279

[3] Zain ME (2011) Impact of mycotoxins on humans and animals. J Saudi Chem Soc 15: 129-144.

[4] Egbuna C, Amadi CN, Patrick-Iwuanyanwu KC, Ezzat SM, Awuchi CG, Ugonwa PO, Orisakwe OE. Emerging Pollutants in Nigeria: A Systematic Review (2021). Environ Toxicol Pharmacol. 2021 Mar 20:103638. doi: 10.1016/j.etap.2021.103638. Epub ahead of print. PMID: 33757839.

[5] Freire FDCO, da Rocha MEB (2016) Impact of mycotoxins on human health. In: Mérillon JM, Ramawat K. (Eds), *Fungal metabolites. Reference Series in Phytochemistry.* Springer, Cham, 239-261.

[6] Umesha S, Manukumar HMG, Chandrasekhar B, et al. (2017) Aflatoxins and food pathogens: impact of biologically active aflatoxins and their control strategies. J Sci Food Agric 97: 1698– 1707.

[7] Alshannaq A, Yu JH (2017) Occurrence, toxicity, and analysis of major mycotoxins in food. Int J Environ Res Public Health 14: 632.

[8] Awuchi C G, I O Amagwula, P Priya, R Kumar, U Yezdani, M G Khan (2020a). Aflatoxins In Foods And Feeds: A Review On Health Implications, Detection, And Control. Bulletin of Environment, Pharmacology, and Life Sciences, 9(9): 149-155

[9] Awuchi, CG; Owuamanam, IC; Ogueke, CC; Hannington T (2020b). The Impacts of Mycotoxins on the Proximate Composition and Functional Properties of Grains. European Academic Research. 8 (2): 1024-1071.

[10] Reverberi M, Ricelli A, Zjalic S, et al. (2010) Natural functions of mycotoxins and control of their biosynthesis in fungi. Appl Microbiol Biotechnol 87: 899-911.

[11] Benkerroum N (2016) Mycotoxins in dairy products: A review. Int Dairy J 62:63-75.

[12] Perdoncini M, Sereia M, Scopel F, et al. (2019) Growth of fungal cells and the production of mycotoxins. Cell growth. DOI: 10.5772/intechopen.86533.

[13] Kochiieru Y, Mankevičienė A, Cesevičienė J, et al. (2020) The influence of harvesting time and meteorological conditions on the occurrence of Fusarium species and mycotoxin contamination of spring cereals. J Sci Food Agric 100: 2999-3006.

[14] Richard JL, Payne GA, Desjardins AE, et al. (2003) Mycotoxins: Risks in plant, animal and human systems. CAST Task Force Rep 139: 101-103.

[15] Magan, N., Medina, A., Aldred, D. 2011. Possible climate-change effects on mycotoxin contamination of food crops pre- and postharvest. Plant Pathology, 60, 150-163.

[16] Paterson RRM, Lima N (2011). Further mycotoxin effects from climate

change. Food Research International, 44, 2555-2566.

[17] Medina, A., Schmidt-Heydt, M., Cárdenas-Chávez, D. L., Parra, R., Geisen, R., Magan, N. 2013. Integrating toxin gene expression, growth and fumonisin B1 and B2 production by a strain of Fusarium verticillioides under different environmental factors. Journal of the Royal Society Interface, 10.

[18] Popovski S., Celar F. A. (2013). The impact of environmental factors on the infection of cereals with Fusarium species and mycotoxin production - A review. Acta Agriculturae Slovenica, 101, 105-116.

[19] Awuchi, Chinaza Godswill; Awuchi, Chibueze Gospel; Ukpe, Aniefiok Emem; Asoegwu, Chisom Rachael; Uyo, Chijioke Ndukwe; Ngoka, Kelechi Enyinnaya (2020c). Environmental Impacts of Food and Agricultural Production: A Systematic Review. European Academic Research. 8 (2): 1120 – 1135.

[20] Biomin. Mycotoxin formation/ fungal growth. Available at http://www.mycotoxins.info/mycotoxins/mycotoxin-formation/. Accessed June 20, 2021.

[21] Smith MC, Madec S, Coton E, et al. (2016) Natural co-occurrence of mycotoxins in foods and feeds and their in vitro combined toxicological effects. Toxins (Basel) 8: 94.

[22] Leslie JF, Bandyopadhyay R, Visconti A (2008) *Mycotoxins: Detection methods, management, public health and agricultural trade*. Wallingford: CAB International.

[23] Mannaa M, Kim KD (2017a) Influence of temperature and water activity on deleterious fungi and mycotoxin production during grain storage. Mycobiology 45: 240-254.

[24] Joubrane K, Mnayer D, El Khoury A, et al. (2020) Co-occurrence of Aflatoxin B1 (AFB1) and Ochratoxin A (OTA) in Lebanese stored wheat. J Food Prot 83: 1547-1552.

[25] Thanushree MP, Sailendri D, Yoha KS, et al. (2019) Mycotoxin contamination in food: An exposition on spices. Trends Food Sci Technol 93: 69-80.

[26] Grigoryan KM, Hakobyan LL (2015) Effect of water activity, pH. and temperature on contamination level of dried vine fruit by filamentous fungi during storage. *Proc of the Yerevan State Univ Chemistry and Biology* 23-28.

[27] Lahouar A, Marin S, Crespo-sempere A, et al. (2016) Effects of temperature, water activity and incubation time on fungal growth and aflatoxin B1 production by toxinogenic Aspergillus flavus isolates on sorghum seeds. Rev Argent Microbiol 48: 78-85.

[28] Kamil OH, Lupuliasa D, Draganescu D, et al. (2011) Interrelations of drying heat and survival of different fungal spores within the tablets formulation. Stud Univ Vasile Goldiş 21: 339-342.

[29] Greeff-laubscher MR, Beukes I, Marais GJ (2019) Mycotoxin production by three different toxigenic fungi genera on formulated abalone feed and the effect of an aquatic environment on fumonisins. Mycology 11: 105-117.

[30] Vylkova S (2017). Environmental pH modulation by pathogenic fungi as a strategy to conquer the host. PLoS Pathog 13: e1006149.

[31] Wang YB, Zhang WG, Fu LL (2017) *Food Spoilage Microorganisms: Ecology and Control*. CRC Press.

[32] Brzonkalik K, Hümmer D, Syldatk C, et al. (2012). Influence of pH and carbon to nitrogen ratio on

mycotoxin production by Alternaria alternata in submerged cultivation. AMB Express 2: 28.

[33] Kokkonen M, Jestoi M, Rizzo A (2005). The effect of substrate on mycotoxin production of selected Penicillium strains. Int J Food Microbiol 99: 207-214.

[34] Duran R, Cary JW, Calvo AM (2010). Role of the osmotic stress regulatory pathway in morphogenesis and secondary metabolism in filamentous fungi. Toxins (Basel) 2: 367-381.

[35] Hamad H, Mehmet A, Ismael H, et al. (2015) The effect of some sugars on the growth of Aspergillus niger. Kahramanmaraş Sütçü İmam Üniversitesi Doğa Bilim Derg 17: 7.

[36] Liu J, Sun LH, Zhang NY, et al. (2016) Effects of nutrients in substrates of different grains on Aflatoxin B 1 production by Aspergillus flavus. Biomed Res Int 2016: 7232858.

[37] Uppala SS, Bowen KL, Woods FM (2013) Pre-harvest aflatoxin contamination and soluble sugars of peanut. Peanut Sci 40: 40-51.

[38] Van der Fels-Klerx HJ (Ine), Liu C, Battilani P (2016) Modelling climate change impacts on mycotoxin contamination. World Mycotoxin J 9: 1-10.

[39] Medina A, Rodríguez A, Magan N (2015) Climate change and mycotoxigenic fungi: Impacts on mycotoxin production. Curr Opin Food Sci 5: 99-140.

[40] Miraglia M, De Santis B, Brera C (2008) Climate change: Implications for mycotoxin contamination of foods. J Biotechnol 136: S715.

[41] Battilani P, Rossi V, Giorni P, et al. (2012) Modelling, predicting and mapping the emergence of aflatoxins in cereals in the EU due to climate change. Scientific report submitted to EFSA.

[42] Medina A, Akbar A, Baazeem A, et al. (2017) Climate change, food security and mycotoxins: Do we know enough? Fungal Biol Rev 31: 143-154.

[43] Valencia-Quintana R, Milić M, Jakšić D, et al. (2020) Environmental research and public health review environment changes, Aflatoxins, and health issues, a Review. Int J Environ Res Public Health 17: 7850.

[44] Moses JA, Jayas DS, Alagusundaram K (2015) Climate change and its implications on stored food grains. Agric Res 4: 21-30.

[45] Joubrane K, El Khoury A, Lteif R, et al. (2011) Occurrence of aflatoxin B1 and ochratoxin A in Lebanese cultivated wheat. Mycotoxin Res 27: 249-257.

[46] Mannaa M, Kim KD (2017b) Control strategies for deleterious grain fungi and mycotoxin production from preharvest to postharvest stages of cereal crops: A Review. Life Sci Nat Resour Res 25: 13-27.

[47] Mahuku G, Nzioki HS, Mutegi C, et al. (2019) Pre-harvest management is a critical practice for minimizing a fl atoxin contamination of maize. Food Control 96: 219-226.

[48] Munkvold G (2014) Crop management practices to minimize the risk of mycotoxins contamination in temperate-zone maize. In: Leslie JF, Logrieco AF. (Eds), *Mycotoxin Reduction in Grain Chain.* 59-77.

[49] Rose LJ, Okoth S, Flett BC, et al. (2019) Preharvest management strategies and their impact on mycotoxigenic fungi and associated mycotoxins. *Fungi and Mycotoxins*. Rijeka: IntechOpen.

[50] Nganchamung T, Robson M (2017) Chemical fertilizer use and acute health effects among chili farmers in Ubon Ratchathani province, Thailand. *J Heal Res* 31.

[51] Reboud X, Eychenne N, Délos M, et al. (2016) Withdrawal of maize protection by herbicides and insecticides increases mycotoxins contamination near maximum thresholds. Agron Sustain Dev 36: 43.

[52] Rodríguez A, Rodríguez M, Andrade MJ, et al. (2015) Detection of filamentous fungi in foods. Curr Opin Food Sci 5: 36-42.

[53] Nicolopoulou-Stamati P, Maipas S, Kotampasi C, et al. (2016) Chemical pesticides and human health: The urgent need for a new concept in agriculture. Front public Heal 4: 148.

[54] Kagot V, Okoth S, Boevre MD, et al. (2019) Biocontrol of aspergillus and fusarium mycotoxins in Africa: benefits and limitations. Toxins (Basel) 11: 109.

[55] Dorner JW (2004) Biological control of Aflatoxin crop contamination. J Toxicol - Toxin Rev 23: 425-450.

[56] Awuchi CG, Amagwula IO (2020). The Biochemistry, Toxicology, and Uses of the Pharmacologically Active Phytochemicals: Alkaloids, Terpenes, Polyphenols, and Glycosides. Merit Research Journal of Food Science and Technology, 5(1): 006-021, DOI: 10.5281/zenodo.3967809

[57] Wambacq E, Vanhoutte I, Audenaert K, et al. (2016) Occurrence, prevention and remediation of toxigenic fungi and mycotoxins in silage: a review. J Sci Food Agric 96: 2284-2302.

[58] Magan N, Olsen M (2004) *Mycotoxins in food detection and control*. Cambridge: CRC.

[59] Karlovsky P, Suman M, Berthiller F, et al. (2016) Impact of food processing and detoxification treatments on mycotoxin contamination. Mycotoxin Res 32: 179-205.

[60] Hojnik N, Cvelbar U, Tavčar-Kalcher G, et al. (2017) Mycotoxin decontamination of food : Cold " classic " decontamination. Toxins (Basel) 9: 151.

[61] Awuchi CG, Twinomuhwezi H, Igwe VS, Amagwula IO (2020d). Food Additives and Food Preservatives for Domestic and Industrial Food Applications. Journal of Animal Health, 2 (1): 1 – 16.

[62] Pleadin J, Frece J, Markov K (2019) Mycotoxins in food and feed. Adv Food Nutr Res 89: 297– 345.

[63] Deng LZ, Sutar PP, Mujumdar AS, et al. (2021) Thermal decontamination technologies for microorganisms and mycotoxins in low-moisture foods. Annu Rev Food Sci Technol 12.

[64] Pankaj SK, Shi H, Keener KM (2018) A review of novel physical and chemical decontamination technologies for a flatoxin in food. Trends Food Sci Technol 71: 73-83.

[65] Awuchi CG, Owuamanam IC, Ogueke CC, Igwe VS (2019). Evaluation of Patulin Levels and impacts on the Physical Characteristics of Grains. International Journal of Advanced Academic Research, 5 (4); 10 – 25.

[66] Misra NN, Yadav B, Roopesh MS, et al. (2019) Cold plasma for effective fungal and Mycotoxin control in foods: Mechanisms, inactivation effects, and applications. Compr Rev food Sci Food Saf 18: 106-120.

[67] Lee GJ, Sim GB, Choi EH, et al. (2015) Optical and structural properties of plasma-treated Cordyceps bassiana spores as studied by circular dichroism,

absorption, and fluorescence spectroscopy. J Appl Phys 117: 023303.

[68] Dasan BG, Boyaci IH, Mutlu M (2017) Nonthermal plasma treatment of Aspergillus spp. spores on hazelnuts in an atmospheric pressure fluidized bed plasma system: Impact of process parameters and surveillance of the residual viability of spores. J Food Eng 196: 139-149.

[69] Deng LZ, Mujumdar A, Pan ZL, et al. (2019) Emerging chemical and physical disinfection technologies of fruits and vegetables: a comprehensive review. Crit Rev Food Sci Nutr 60: 1-28.

[70] Ouf SA, Basher AH, Mohamed AH (2014) Inhibitory effect of double atmospheric pressure argon cold plasma on spores and mycotoxin production of Aspergillus niger contaminating date palm fruits. J Sci Food Agric 95: 204-210.

[71] Gavahian M, Cullen PJ (2020) Cold plasma as an emerging technique for mycotoxin-free food: Efficacy, mechanisms, and trends. Food Rev Int 36: 193-214.

[72] Murugesan P, Brunda DK, Moses JA, et al. (2020) Photolytic and photocatalytic detoxification of mycotoxins in foods. Food Control 123: 107748.

[73] Mao J, He B, Zhang L, et al. (2016) A structure identification and toxicity assessment of the degradation products of aflatoxin B_1 in peanut Oil under UV irradiation. Toxins (Basel) 8: 332.

[74] Ibarz R, Garvín A, Azuara E, et al. (2015) Modelling of ochratoxin A photo-degradation by a UV multi-wavelength emitting lamp. LWT - Food Sci Technol 61: 385-392.

[75] Magzoub RAM, Yassin AAA, Abdel-Rahim AM, et al. (2018) Photocatalytic detoxification of aflatoxins in Sudanese peanut oil using immobilized titanium dioxide. Food Control 95: 206-214.

[76] Colovic R, Puvača N, Cheli F, et al. (2019) Decontamination of mycotoxin-contaminated feedstuffs and compound feed. Toxins (Basel) 11: 617.

[77] Patriarca A, Fernandez Pinto V (2017) Prevalence of mycotoxins in foods and decontamination. Curr Opin Food Sci 14: 50-60.

Chapter 3

Fungal Growth and Pathology

Ozlem Gulmez and Ozlem Baris

Abstract

Fungi, an important group with a wide variety of species, shows spectacular development with their unique cell structures. Fungi survive in many different ecosystems with their reproductive abilities and metabolic features. Thanks to wide temperature and pH tolerances, fungi develop on organic and inorganic materials in all ecosystems they are in and maintain the existence of ecosystems by taking part in many cycles. However, examples of pathogens are also available. They are a group of organisms that are environmentally important, such as saprophytes and mutualists, but are pathogens for animals, especially plants. Fungi basically have two different cell structures: yeast, and molds. But some fungi have both of these structures. Depending on the temperature of the environment they are in, they can be found in yeast or mold structures, and fungi with this feature are called dimorphic fungi. Whether it is yeast, mold, or dimorphic fungi, they use their enzymes with high activity to benefit from the nutrients in the environment. Fungi can be easily grown in natural and synthetic media. Yeast can reproduce rapidly with their single-celled structure, while molds and mushrooms are very successful with their hyphae structures.

Keywords: fungal growth, pathology, reproduction

1. Introduction

Fungi are eukaryotic organisms thought to have about 4 million species [1]. Although the cell structures of fungi are similar to other eukaryotic cells, they differ from other cells by the presence of ergosterol in their cell membrane and chitin in their cell walls. Cell cytoplasm contains higher concentrations of salt and sugar than other eukaryotes. This regulates cell homeostasis and regulates the exchange of substances. Except for yeasts, most fungi have microscopic structures called hyphae, and these come together to form visible structures called mycelium. The hyphae have apical growth. In the apical growing parts of the hyphae, there are secretory vesicles called "Spitzenkörper" and Wooronin body organs that act as peroxisomes. Because of these properties of hyphae, fungi can live where other eukaryotic cells do not and can use various substrates [2, 3]. Fungi take part in many degradations, transformation, and cycle events in nature. Although they are heterotrophic creatures, they can survive as saprophytic, mutualistic, and parasitic. The fact, that they are found in all parts of the world and live in different environments is due to the superior reproductive abilities of fungi [4–6].

1.1 Fungal reproduction

Fungi can reproduce sexually and asexually. Asexual reproduction of fungi is carried out by vegetative reproduction through hyphae or by the spores they produce. Sexual reproduction is; they form a diploid nucleus with the union of haploid spores, and the cycle continues with the germination of this nucleus. In fungi, asexual reproduction takes place more than sexual reproduction. This event increases the adaptive power of fungi and prevents the accumulation of any harmful mutations that may occur and their transmission from generation to generation. In addition, their chances of survival and competitive advantage are ensured [4, 5, 7].

Sexual reproduction in fungi takes place in three stages. In the first stage; haploid cells fuse, this is called plasmogamy. In the second stage; the fusion of two haploid nuclei, this event is karyogamy. The third stage is; the resulting diploid cells undergo meiosis to form haploid cells, and the cycle continues in this way [8]. These stages are summarized in **Figure 1**.

The association or non-union of haploid cells is determined by DNA. The sex of haploid cells is determined by a specific gene region in fungi. This region is known as the mating-type locus and is abbreviated MAT. MATs genetically determine the mating identity of fungi and stimulate the secretion of pheromones. Secreted pheromones provide communication between fungi and realize sexual intercourse [4, 9–11].

1.2 Growth and development needs of fungi

Fungi, like every living thing, need energy and food sources to complete their development and life cycles after sexual and asexual reproduction. These food sources are carbon, nitrogen, vitamins, and minerals. They also need suitable environmental conditions (such as pH, temperature, humidity, oxygen) to grow and develop [12–14].

Fungi can consume vegetable and animal carbon sources thanks to their hydrolytic enzymes. They can use monosaccharides and polysaccharides such as glucose, fructose, chitin, cellulose, hemicellulose, and lignin [15, 16]. Like all living things,

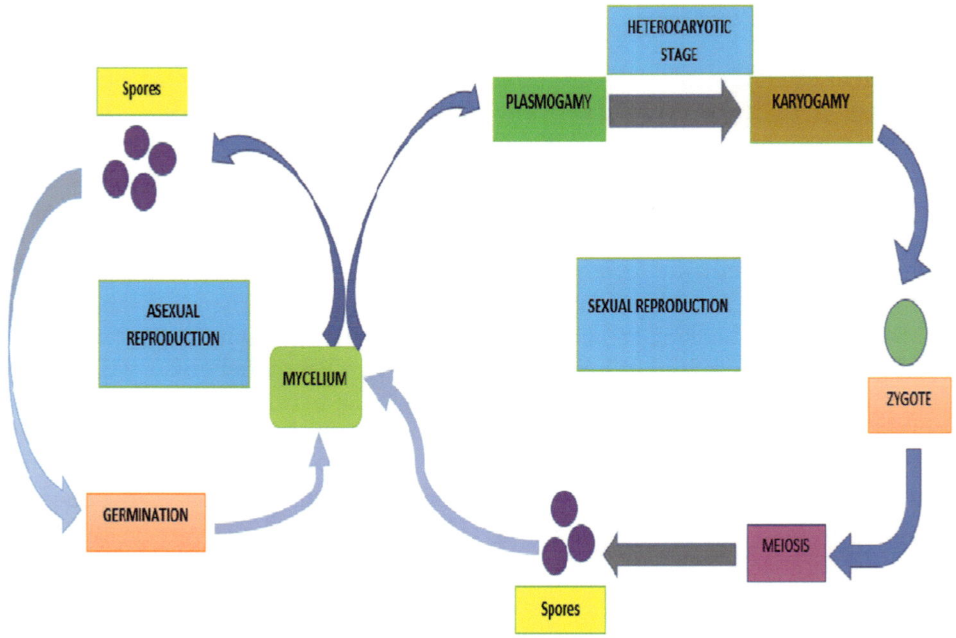

Figure 1.
Fungal reproduction.

fungi need a nitrogen source for their growth and development, and fungi can metabolize many different nitrogen sources. Especially ammonium and glutamine are the first nitrogen sources they use. In addition, they can easily use other nitrogen sources [17].

Vitamins are cofactors of enzymes and growth factors of many organisms. Fungi need vitamins for their growth and development. Some of these vitamins are; thiamine, biotin, riboflavin, nicotinic acid, vitamin K and pantothenic acid [18].

Like many microorganisms, fungi can survive in varying environmental conditions and under various stress factors. They can survive and reproduce in extreme environments, such as the poles, in extremely cold regions, and in extremely hot regions such as deserts. Fungi are generally; grow better in warm, acidic, and aerobic environments, but they can survive in cold, alkaline, and anaerobic environments. Although the growth temperatures of the fungi are quite wide, the best growth is seen at 25°C. Fungi that live under the temperature at which they develop optimally are called psychrotolerant, and fungi that live at temperatures of 40°C and above are called thermotolerant fungi. Fungi that live in or are exposed to temperatures above 40°C can survive by protecting themselves from heat stress by producing heat shock proteins. Fungi can be found in yeast or mold structures depending on the temperature of the environment they are in, and fungi with this feature are called dimorphic fungi. One of the most important fungi showing this feature is *Histoplasma* sp. If the place where it is located in an environment of 25°C, it develops as a mold with a hyphae structure that can reproduce vegetatively, and as yeast if the environment is 37°C. This feature allows them to survive in different ecosystems and even to continue their generation [19]. The pH value, which is important for the realization of many biochemical reactions in all living things, is one of the important environmental conditions for fungi. Some fungi can survive and even reproduce at pH 1 and 13, which are extreme for many organisms. Most fungi survive and reproduce between pH 3 and 10. The optimum pH range is between 5 and 7 [20–22]. In addition, if the environment in which fungi are found in alkaline, they can achieve maximum growth by converting the pH of the environment to the optimum growth pH with the organic acids they secrete. Abiotic stress factors such as water, UV, and heavy metals affect the development of fungi as well as all organisms. In particular, UV-B radiation, which is more biologically harmful, negatively affects the growth and development of fungi [23].

Fungi are among the largest and most diverse groups of eukaryotic organisms. Because of their complex gene structure, the enzymes they produce, and their ability to use many different carbon sources, they, directly and indirectly, affect human life. They have been used for centuries as a food source and in the production process of many biotechnological products. Today, fungi are used in various fields such as antibiotics, enzyme technology, drug production, pigment production [24–26]. Although fungi are necessary for the survival of life on earth, they cause serious problems in most organisms. Fungi cause disease in humans, animals, and plants and cause the death of these organisms [27, 28]. Fungi infect many organisms with the secondary metabolites and mycotoxins they produce, even cause their extinction [29].

1.3 Pathogenic fungi

A fungal kingdom is a group that contains the most and most harmful plant pathogens. By infecting all tissues and organs of plants, they damage many herbaceous and woody plants with high economic value and nutritional properties. In cultivated plants such as corn, wheat, sugar beet, potato, banana; causes great harm to farmers by causing diseases such as root rot, wilt, stem softness, gall and rust.

Pathogen fungus	Plant	Damage
Heterobasidion parviporum	Picea	Root rot
Heterobasidion annosum	Abies	Root rot
Ustilago maydis	Maize	Gall
Sporisorium reilianum	Maize, Sugar beet	Ear rot
Puccinia graminis f. sp. *tritici*	Wheat, barley	Black rust
Melampsora lini	Flaxseed	Red rust
Fusarium oxysporium	Tomato, pepper, watermelon	Root rot
Alternaria sp.	Chickpeas, carrots, olives, apples	Pallidness
Apergillus niger	Ginger	Aflatoxin
Leptosphaeria maculans	Crucifers	Blackleg
Podosphaera plantaginis	Banana	Mildew
Rhizoctonia solani	Soybean	Root rot
Dothistroma septosporum	Pinus	Red band needle blight
Lecanosticta acicola	Pinus	Brown spot needle blight,
Phytophthora infestans	Tomato, potato, pepper eggplant	Mildew
Verticilum sp.	Plants	Vascular pallor

Table 1.
Some plant pathogenic fungi and infecting plants.

Pathogen fungus	Organisms	Damage
Fusarium sp.	Horse	Encephalomalacia
Aspergillus sp.	Human	Liver cancer
Cladophialophora bantiana	Human	Brain tissue loss
Candida auris	Human	Infection in the blood
Cryptococcus sp.	Human	Infection in the lung
Batrachochytrium dendrobatidis	Amphibian	
Laboulbenia formicarum	Ant	
Aspergillus fumigatus	Bird	Lung infection
Fusarium, Ochroconis, Exophiala, Scytalidium, Plectosporium, and *Acremonium.*	Fish shellfish	
Microsporum canis and *Sporothrix brasiliensis*	Cat	

Table 2.
Diseases are caused by some pathogenic fungi in humans and animals.

They also develop in stored grains and cause product loss. Also; high woody plants are infected by white rot and brown rot fungi, resulting in tissue deterioration and plant death. Plant pathogen fungi can reproduce both sexually and asexually in host plants [30–33]. **Table 1** shows the plants that some fungi cause disease.

Fungi cause infections not only in plants but also in humans and animals. Bees, insects, frogs, fish, and corals are some organisms affected by fungal infections. Fungi enter the body from the outer shells, trachea, and skin of these creatures and cause the death of these creatures. Fungal diseases have killed more than 1.6 million people annually. It is thought that pandemics caused by fungi may occur with global warming and climate change [34]. Because the stress tolerance and adaptation abilities of the fungi are very high, they have destroyed their existence on earth by infecting many different organism groups (**Table 2**). In humans, they cause skin infection, lung infection, and intestinal infection. They also cause diseases in animals and humans by reproducing sexually and asexually [35, 36].

2. Conclusions

Apart from its use as food; the fungi which we use in the production of drugs, antibiotics, anticarcinogenic substances, pigments, alcohol, and biofuels, are indispensable elements for the continuation of our lives. As we stated in this publication, their high reproductive capacity and ability to survive in extreme conditions provide fungi with a competitive advantage and advantage over other organisms. These abilities give it the capability to live in the plant, animal, and human tissue-organs. If the development and growth demand of fungi, whether pathogenic or not, are known, we can make the most of these organisms and prevent the development of fungi that cause disease. This study will enable us to get to know fungi a little more closely and will enable us to take precautions against these organisms.

Conflict of interest

The authors do not declare any conflict of interest.

Author details

Ozlem Gulmez* and Ozlem Baris
Faculty of Science, Biology Department, Ataturk University, Erzurum, Turkey

*Address all correspondence to: ozlmg90@gmail.com

References

[1] Nilsson RH, Larsson KH, Taylor AFS, Bengtsson-Palme J, Jeppesen TS, Schigel D, et al. The UNITE database for molecular identification of fungi: Handling dark taxa and parallel taxonomic classifications. Nucleic Acids Research. 2019;**47**(D1):D259-D264

[2] Money NP. Fungi and biotechnology. In: The Fungi. Oxford, OH, USA: Academic Press; 2016. pp. 401-424

[3] Raghukumar S. Fungi: Characteristics and classification. In: Fungi in Coastal and Oceanic Marine Ecosystems. Cham: Springer; 2017. pp. 1-15

[4] Peraza-Reyes L, Malagnac F. 16 sexual development in Fungi. In: Growth, Differentiation and Sexuality. Cham: Springer; 2016. pp. 407-455

[5] Dyer PS, Inderbitzin P, Debuchy R. 14 mating-type structure, function, regulation and evolution in the pezizomycotina. In: Growth, Differentiation and Sexuality. Cham: Springer; 2016. pp. 351-385

[6] Li D, Tang Y, Lin J, Cai W. Methods for genetic transformation of filamentous fungi. Microbial Cell Factories. 2017;**16**(1):1-13

[7] Döll K, Chatterjee S, Scheu S, Karlovsk P, Rohlfs M. Fungal metabolic plasticity and sexual development mediate induced resistance to arthropod fungivory. Proceedings of the Royal Society B: Biological Sciences. 2013;**280**(1771):20131219

[8] Wallen RM, Perlin MH. An overview of the function and maintenance of sexual reproduction in dikaryotic fungi. Frontiers in Microbiology. 2018;**9**:503

[9] Lee SC, Heitman J. Sex in the Mucoralean fungi. Mycoses. 2014;**57**:18-24

[10] Ni M, Feretzaki M, Sun S, Wang X, Heitman J. Sex in fungi. Annual Review of Genetics. 2011;**45**:405-430

[11] Nieuwenhuis BP, Billiard S, Vuilleumier S, Petit E, Hood ME, Giraud T. Evolution of uni-and bifactorial sexual compatibility systems in fungi. Heredity. 2013;**111**(6):445-455

[12] Basu S, Bose C, Ojha N, Das N, Das J, Pal M, et al. Evolution of bacterial and fungal growth media. Bioinformation. 2015;**11**(4):182

[13] Wongjiratthiti A, Yottakot S. Utilization of local crops as alternative media for fungal growth. Pertanika Journal of Tropical Agricultural Science. 2017;**40**:295-304

[14] Sawiphak S, Wongjiratthiti A, Saengprasan C. Dioscorea alata as alternative culture media for fungal cultivation and biomass production. Pertanika Journal of Tropical Agricultural Science. 2021;**44**(2):327, 336

[15] Homaei A, Ghanbarzadeh M, Monsef F. Biochemical features and kinetic properties of α-amylases from marine organisms. International Journal of Biological Macromolecules. 2016;**83**:306-314

[16] Pawłowska J, Okrasińska A, Kisło K, Aleksandrzak-Piekarczyk T, Szatraj K, Dolatabadi S, et al. Carbon assimilation profiles of mucoralean fungi show their metabolic versatility. Scientific Reports. 2019;**9**(1):1-8

[17] Tudzynski B. Nitrogen regulation of fungal secondary metabolism in fungi. Frontiers in Microbiology. 2014;**5**:656

[18] Tanner RS. Cultivation of bacteria and fungi. In: Manual of Environmental Microbiology. Third ed. Wasington, DC, USA: American Society of Microbiology; 2007. pp. 69-78

[19] Rodriguez L, Voorhies M, Gilmore S, Beyhan S, Myint A, Sil A. Opposing signaling pathways regulate morphology in response to temperature in the fungal pathogen Histoplasma capsulatum. PLoS Biology. 2019;**17**(9):e3000168

[20] Peñalva MA, Tilburn J, Bignell E, Arst HN Jr. Ambient pH gene regulation in fungi: Making connections. Trends in Microbiology. 2008;**16**(6):291-300

[21] Zhang L, Song X, Shao X, Wu Y, Zhang X, Wang S, et al. Lead immobilization assisted by fungal decomposition of organophosphate under various pH values. Scientific Reports. 2019;**9**(1):1-9

[22] Alves TP, Triques CC, da Silva EA, Fagundes-Klen MR, SDM H. Multi-enzymatic recovery of fungal cellulases (*Aspergillus niger*) through solid-state fermentation of sugarcane bagasse. The Canadian Journal of Chemical Engineering. 2021;**1**:11. DOI: 10.1002/cjce.24292

[23] Hughes KA, Lawley B, Newsha KK. Solar UV-B radiation inhibits the growth of Antarctic terrestrial fungi. Applied and Environmental Microbiology. 2003;**69**(3):1488-1491

[24] Solomon L, Tomii VP, Dick AAA. Importance of fungi in the petroleum, agro-allied, agriculture and pharmaceutical industries. New York Science Journal. 2019;**12**:8-15

[25] Li Y, Jin W, Mu C, Cheng Y, Zh W. Indigenously associated methanogens intensified the metabolism in hydrogenosomes of anaerobic fungi with xylose as substrate. Journal of Basic Microbiology. 2017;**57**(11):933-940

[26] Matassa S, Boon N, Pikaa I, Verstraet W. Microbial protein: Future sustainable food supply route with low environmental footprint. Microbial Biotechnology. 2016;**9**(5):568-575

[27] Hyde KD, Norphanphou C, Chen J, Dissanayake AJ, Doilom M, Hongsanan S, et al. Thailand's amazing diversity: Up to 96% of fungi in northern Thailand may be novel. Fungal Diversity. 2018;**93**(1):215-239

[28] Hyde KD, Tennakoon DS, Jeewon R, Bhat DJ, Maharachchikumbura SS, Rossi W, et al. Fungal diversity notes 1036-1150: Taxonomic and phylogenetic contributions on genera and species of fungal taxa. Fungal Diversity. 2019; **96**(1):1-242

[29] Gruber-Dorninger C, Jenkins T, Schatzmayr G. Global mycotoxin occurrence in feed: A ten-year survey. Toxins. 2019;**11**(7):375

[30] Javelle M, Vernoud V, Rogowsky PM, Ingram GC. Epidermis: The formation and functions of a fundamental plant tissue. New Phytologist. 2011;**189**(1):17-39

[31] Menardo F, Praz CR, Wyder S, Ben-David R, Bourras S, Matsumae H, et al. Hybridization of powdery mildew strains gives rise to pathogens on novel agricultural crop species. Nature Genetics. 2016;**48**(2):201-205

[32] Möller M, Stukenbrock EH. Evolution and genome architecture in fungal plant pathogens. Nature Reviews Microbiology. 2017;**15**(12):756-771

[33] Casadevall A. Don't forget the fungi when considering global catastrophic biorisks. Health Security. 2017;**15**(4): 341-342

[34] Gostinčar C, Zajc J, Lenassi M, Plemenitaš A, De Hoog S, Al-Hatmi AM, et al. Fungi between extremotolerance and opportunistic pathogenicity on humans. Fungal Diversity. 2018;**93**(1):195-213

[35] Seyedmousavi S, Bosco SDM, De Hoog S, Ebel F, Elad D, Gomes RR, et al. Fungal infections in animals: A

patchwork of different situations. Medical Mycology. 2018; **56**(suppl_1):S165-S187

[36] Almeida F, Rodrigues ML, Coelho C. The still underestimated problem of fungal diseases worldwide. Frontiers in Microbiology. 2019;**10**:214

Section 2

Fungal Behaviour

Chapter 4

Fungal Immunology: Mechanisms of Host Innate Immune Recognition and Evasion by Pathogenic Fungi

Faisal Rasheed Anjum, Sidra Anam, Muhammad Luqman, Ameena A. AL-surhanee, Abdullah F. Shater, Muhammad Wasim Usmani, Sajjad ur Rahman, Muhammad Sohail Sajid, Farzana Rizvi and Muhammad Zulqarnain Shakir

Abstract

For a fungal pathogen to successfully infect, colonize and spread inside a susceptible host, it must have overcome the host immune responses. The early recognition of the fungal pathogen-associated molecular patterns (PAMPS) by the host's pattern recognition receptors (PRRs) results in the establishment of anti-fungal immunity. Although, our immune system has evolved several processes to combat these pathogens both at the innate and adaptive immune levels. These organisms have developed various escape strategies to evade the recognition by the host's innate immune components and thus interfering with host immune mechanisms. In this chapter, we will summarize the major PRRs involved in sensing fungal PAMPS and most importantly the fungal tactics to escape the host's innate immune surveillance and protective mechanisms.

Keywords: PAMPs, PRRs, innate immunity, escape mechanisms, pathogenic fungi

1. Introduction

Pathogenic fungi are an important cause of morbidity and mortality in humans particularly in immune-compromised individuals [1, 2]. The most common risk factors for the increased incidence of fungal infections in immunocompromised individuals are cancer therapy, use of corticosteroids and neutropenia [3–5]. Sporadic occurrence of fungal infections has also been described in immunocompetent individuals that have undergone any traumatic inoculation such as the use of catheters or surgeries [6, 7]. Fungal pathogens show a considerable variation in their biology and disease pathogenesis and may include opportunistic fungi, i.e., *Aspergillus fumigatus* (*A. fumigatus*), and *Fusarium spp.* as well as some commensals such as *Candida albicans (C. albicans)*.

The human innate immune system is the first line of defense and plays a pivotal role in the body's defense on confrontation to the invading pathogens. One of the fundamental responses towards the infectious agents including fungi is the inflammatory response that is launched immediately by the host body following an immunological insult. This inflammatory response drives the antigens specific adaptive immune response such as activation of antigen-specific lymphocytes against the invading pathogens. The innate immune system recognizes a particular set of conserved surface molecules exhibited by the pathogens called pathogens-associated molecular patterns (PAMPs). Host cell pattern recognition receptors (PRRs) detect microbial PAMPs and trigger the intracellular signaling pathways that lead to the production of cytokines, reactive oxygen species (ROS), reactive nitrogen species (RNS), and lipid mediators [8, 9]. Toll-like receptors (TLRs) and C-type lectin receptors (CLRs) are the most common PRRs characterized for the detection of fungal PAMPs. Microbial detection by these PRRs results in a cascade of signaling events that eventually result in the production of inflammatory mediators, phagocytosis and induction of adaptive immune response [10]. However, fungal pathogens have adapted simple yet innovative strategies to evade and/or counteract the host innate immune responses thus resulting in the establishment of a successful infection inside the host. In the current chapter, we have made a comprehensive understanding of the major innate immune receptors involved in the detection of the fungal pathogens as well as the strategies employed by the pathogenic fungi to evade and therefore enhance their viability inside the host during infection.

2. Innate immune recognition of fungal PAMPs by host PRRs

2.1 Role of TLRs in recognition of pathogenic fungi

TLRs are a family of receptors that share structural homology with the Toll receptor (first described in the *Drosophila*). To date, 13 types of human and 10 types of murine TLRs have been discovered [11–13]. Generally, TLRs are comprised of extracellular and intracellular domains. The extracellular domain is rich in leucine repeats whereas the intracellular domain shares homology with the Toll/IL-1 receptor (TIR) domain. The TIR domain recruits the adapter proteins such as MyD88, TRIF, TRAM, and TIRAP followed by the initiation of intracellular signaling pathways which eventually result in the activation of different transcription factors, i.e., NF-κB, AP-1, IRFs (IRF3/7), and MAP kinases. These transcription factors lead to the expression of cytokines and co-stimulatory molecules [11].

Both TLR2 and TLR4 are involved in the innate immune recognition of fungal PAMPs (**Table 1**) (**Figure 1**) [10]. TLR4 has been described to recognize the fungal-derived mannans. Recognition of *Saccharomyces cerevisiae* (*S. cerevisiae*) and *C. albicans* derived mannans by human monocytes have been attributed to a mechanism dependent on TLR4 and CD14 with Lipopolysaccharide Binding Protein (LBP) amplifying this mechanism [14]. Further investigation in this regard reveals that recognition of mannans is brought by the cooperation of TLR4 with Mannose Receptor (MR) with TLR4 recognizing the O-linked mannans and MR recognizing the N-linked mannans [15]. TRL4 is required for the innate immune recognition of rhamnomannans isolated from *P. boydii*. Rhamnomannans trigger cytokine production from macrophages *via* TLR4 activation [16]. Similarly, TLR4 also detects glucuronoxylomannans (GXM) from *Cryptococcus neoformans* (*C. neoformans*) suggesting the vital role of TLR4 in innate immune recognition of mannose-containing polysaccharides [17].

Alike TLR4, TLR2 is also involved in the recognition of the fungal molecules. TLR2 triggers the activation of NF-κB and subsequent release of cytokines from the macrophages in response to phospholipomannan (a cell wall lipoglycan isolated from *C. albicans*). On the other hand, both TLR4 and TLR6 respond partially to

PRRs	Fungal pathogen	Fungal PAMPs
1. TLRs		
TLR4	O-linked mannans	*C. albicans*
	Mannans	*Sacchromyces spp.*
	Rhamnomannans	*P. boydii*
	Phospholipomannans	*C. albicans*
	Glucuronuxylomannans	*C. neoformans*
	Unknown/α-, β-glucan, galactomannan	*A. fumigatus*
	Unknown	*P. brasilensis*
TLR2/TLR6	Phospholipomannans	*C. albicans*
	Glucuronuxylomannans	*C. neoformans*
TLR2/TLR1	Glucuronuxylomannans	*C. neoformans*
TLR2	A 1,4 glucans	*P. boydii*
	Unknown	*A. fumigatus* (conidia and hyphae form
2. CLRs		
MR	N-linked mannans	*C. albicans*
	Mannans	*P. Carinii*
	Mannoproteins	*C. neoformans* *A. fumigatus*
	gp43	*P. brasilensis*
Dectin-2	α-mannans	*C. albicans* *A. fumigatus*
Dectin-1	B (1, 3)- glucans	*C. albicans* *A. fumigatus* *Sacchromyces spp.*
DG-SIGN	Galactomannans	*A. fumigatus*
	Mannans	*C. albicans*
	Unknown/surface carbohydrates in extracellular vesicles	*P. brasilensis*
Mincle	Polysaccharides containing α-mannosyl residues	*Malassezia spp.* *C. albican*
CD14	Mannans	*Sacchromyces spp.*
	Unknown	*A. fumigatus*
	A (1,4) glucans	*P. boydii*
3. NNLRs		
NLRP3	Unknown/ β (1,3)-glucans	*C. albicans* *A. fumigatus* *C. neoformans* *P. brasilensis*

Table 1.
List of major PRRs involved in recognition of various fungal-derived PAMPs.

phospholipomannan [18]. Moreover, TLR2 is responsible for the detection of glucogen (i.e., α-1,6-branched α-1,4-glucans) [19]. TLR2/TLR1 and TLR2/TLR6 heterodimers are described to be important receptors in the detection of GXM isolated from the capsules of *C. gatii* and *C. neoformans* [20].

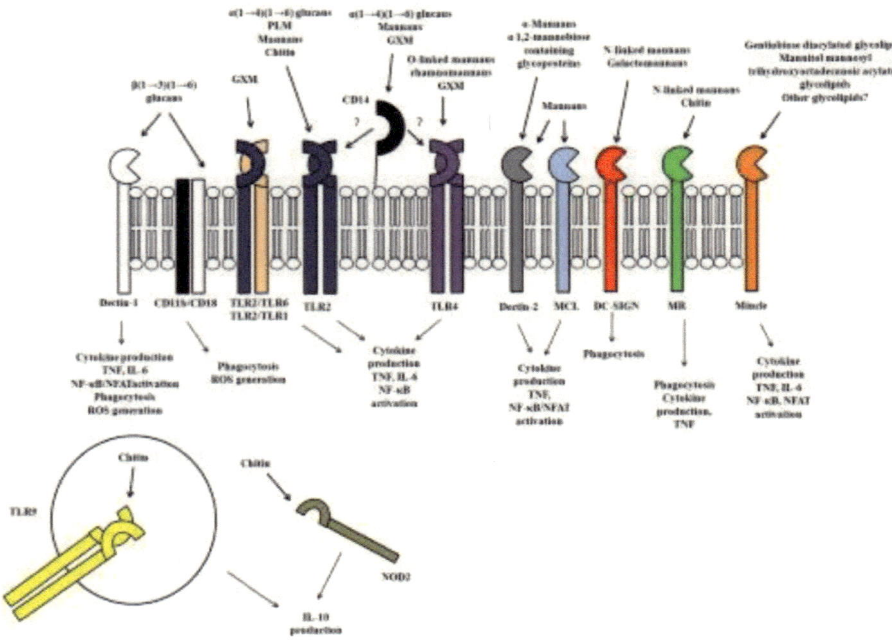

Figure 1.
Major receptors involved in the innate immune recognition of fungal polysaccharides and glycoconjugates (Barreto-Bergter and Figueiredo, 2014).

It has been observed that cytokine production from macrophages and dendritic cells is mediated by TLR2 and CD14 in response to *P. boydii*- derived α-glucans [21]. Polysaccharides extracted from medicinal fungi (*Ganoderma lucidum* and *Cordyceps sinensis*) possess immune-modulatory and anticancer activities. These polysaccharides also trigger cytokine production and B cells activation through TLR2 and TLR4. Although, direct binding of fungal polysaccharides with the TLR2 and TLRs has been described [22]. The exact underlying mechanisms by which TLR2 and TLR4 interact and recognize fungal polysaccharides and other glycoconjugates containing mannose are still poorly defined. Fungal polysaccharides such as α-glucans and mannans are structurally different from the TLR2- and TLR4-prototypical agonists. There could be a possibility that complex fungal polysaccharides or cell wall components may present some uncharacterized glycolipids anchored to their structures that could serve as true TLR2- and/or TLR4- agonists. However, to find these answers, sensitive analytical techniques such as mass spectromery and nuclear magnetic resonance are required in combination with other complementary approaches, i.e., selective inhibition of investigating molecules, use of chemically defined ligands, availability of genetic models deficient in synthesizing the specific fungal molecules. Both TLR4 and TLR2 interact directly with bacterial lipid a and lipopeptides, respectively. Moreover, crystallographic studies of lipid A-TLR4 complex and lipopeptide-TLR1/TLR2 or lipopeptide-TLR6/TLR2 complex have demonstrated a physical interaction between these ligands and their respective receptors [23–25]. It was observed that fatty acid chains present in the bacterial ligands interact and bind with the hydrophobic pockets in the extracellular domains of the receptors complex. This suggests that TLR4 and TLR2 interact with hydrophilic ligands such as fungal polysaccharides in a way that is distinct from that of classical bacterial ligands. Thus, knowing the structural basis for interactions between fungal polysaccharides and TLR2 or TLR4 will be very helpful in understanding how distinct structures, i.e., LPS (lipopolysaccharide), lipopeptides and mammalian endogenous molecules (hyaluronic acid, carboxy alkyl pyrroles, heme) are recognized by these receptors [26].

2.2 Role of CLRs in recognition of pathogenic fungi

CLRs are a group of proteins receptors involved in the detection of fungal glycoconjugates and are characterized by the presence of two motifs; the EPN motif and QPD motifs both of which drive the specificity of CTLRs towards carbohydrate moieties. The EPN motif helps in the recognition of mannose, N-acetylglucosamine, glucose and L-fucose whereas the QPD motif is involved in the recognition of N-acetylgalactosamine and galactose [27–31]. The major CTLRs implicated in the recognition of fungal molecules are Dectin-1, Dectin-2, Mannose receptors (MR), Mincle, DC-SIGN, CD-12, and CD-11b/CD18 (**Table 1**) (**Figure 1**).

2.2.1 Mannose receptor (MR)

The mannose receptor is a type I transmembrane receptor that contains an N-terminal cysteine-rich domain and a type II fibronectin domain. The extracellular component of MR comprises of eight CTLDs (C-type lectin domains) whereas the intracellular component possesses a motif required for endocytic signaling [32]. MR is capable of recognizing fungal PAMPs that contain either mannose, glucose, N-acetylglucosamine [32–34], or sulfated galactose and sulfated N-acetylglucosamine (**Table 1**) (**Figure 1**). Recognition of sulfated glycoconjugates is mediated by the cysteine-rich domain of MR and is independent of CTLDs [35, 36]. On the other hand, recognition of mannose-based fungal PAMPs is dependent on the activity of CTLDs (mostly 4–8) [34]. Several fungal pathogens can be detected by MR including *C. albicans* and *Pneumocystis carinii* [15, 37, 38]. MR is considered a phagocytic receptor due to its involvement in the phagocytosis of pathogenic fungi (**Table 1**) [37, 38]. However, the expression of MR on non-phagocytic cells questions its designation as a professional phagocytic receptor [39]. The role of MR in the recognition and phagocytosis of pathogenic fungi containing mannose ligands is well established [37, 38]. There is a possibility that role of MR as a phagocytic receptor might be cell type-specific.

Besides its role in endocytosis, MR also contributes to the production of cytokines in response to glycoconjugates comprising mannans [15, 40, 41]. On recognizing the *C. albicans* derived mannans, MR induces the release of TNF-α from macrophages [15]. MR is also considered to be involved in sensing of mannosylated glycoconjugates such as those derived from mycobacteria and *Pichia pastoris* indicating that in addition to *C. albicans* derived mannans, MR can also recognize mannosylated glycoconjugates presented from other pathogenic fungi [42, 43]. The mechanism of cytokine induction by MR in response to fungal glycoconjugates is still undefined. MR also contains a short cytoplasmic tail that lacks motifs involved in intracellular signaling and eventually cytokine production. There could be a possibility that MR confers cytokine production in association with other CLRs or TLRs.

2.2.2 Dectin-2

A transmembrane protein that was first characterized in a cell line derived from Langerhans cells. Dectin-2 is comprised of a short cytoplasmic domain and an extracellular domain with CLTD present in the COOH-terminal region. Activation of Dectin-2 by the fungal ligands leads to the production of eicosanoids and cytokines [44–46]. Usually, macrophages, some dendritic cells and IL-6/IL-23-stimulated neutrophils express Dectin-2 receptors [45, 47, 48]. The EPN motif present in the extracellular domain of Dectin-2 recognizes fungal glycoconjugates containing mannose and fucose (**Table 1**) (**Figure 1**) [47]. The binding of Dectin-2 to zymosan requires Ca^{2+}, however, higher concentrations of mannose, fucose, glucose, galactose and N-acetylglucosamine can inhibit this binding. Dectin-2 shows a higher binding affinity towards synthetic carbohydrates that are extensively mannosylated. However, the binding affinity decreases with the decrease in mannosylated residues [49]. Dectin-2 receptor recognizes *C. albicans* by binding to its α-mannans [46, 50]. In case of *Malassezia spp.*, this recognition is mediated through the binding of Dectin-2 with a glycoprotein containing O-linked α-1,2-mannobiose [51]. Besides this, in cooperation with MCL, Dectin-2 also has been described to promote recognition of *C. albicans* hyphae through binding to α-mannans followed by the formation of heterodimer complex. This cooperation between Dectin-2 and MCL results in higher sensitivity in recognizing the *C. albicans* and eventually leads to amplified leukocyte responses towards the fungal pathogens [50].

2.2.3 Dectin-1

Dectin-1 is a type II transmembrane receptor that recognizes and binds to the molecules containing β (1,3)-glucans. It is expressed by many cell types including macrophages, dendritic cells, eosinophils and neutrophils [52–54]. Dectin-1 mediated signaling results in the production of cytokines [55], maturation of dendritic cells [56] and production of ROS [57]. This suggests that Dectin-1 is an important PRR in recognizing β (1,3)-glucans followed by leukocytes activation and induction of adaptive immunity. However, in contrast to many other CTLRs, Dectin-1 binding to β-glucans is not dependent on Ca^{2+} [58, 59]. Dectin-1 possesses higher specificity towards β (1,3)-glucans having β (1,6)-branches. On the other hand, Dectin-1 is unable to bind with mannans, pullulans, β1,6-glucans or β (1,3)/(β1,4)-glucans [58]. In contrast to TLRs which recognize soluble ligands, activation of Dectin-1 is dependent upon its clustering by the β-glucans molecules followed by exclusion of tyrosine phosphatases (CD45 and CD48) and phosphorylation of hemi-ITAM motif present in the cytoplasmic tail of Dectin-1 [60, 61]. The hemi-ITAM recruits Syk kinases and initiates the upstream signaling pathway leading to the activation of NF-κB and NFAT [60, 62]. Usually, alveolar macrophages (AMs), resident peritoneal macrophages and dendritic cells present Dectin-1 dependent responses towards β-glucans while bone marrow-derived macrophages do not. However, Dectin-1 dependent responses can be promoted in non-responding cells such as bone marrow-derived macrophages in the presence of IFN-γ and GM-CSF thus suggesting that responses mediated by Dectin-1 are flexible [63, 64].

2.2.4 Mincle

Also known as CLEC4E is a type II transmembrane protein that was first identified as a macrophage-expressed gene dependent on the activity of the NF-IL6 transcription factor [65]. It is composed of a short cytoplasmic tail with an extracellular domain containing a CLTD. Mincle triggers cell signaling by recruiting the FcRγ chain which leads to the activation of NFAT and NF-κB and eventually induces transcription of cytokines [66]. Like other CTLRs, Mincle also plays an important role in the recognition of fungal molecules (**Table 1**) (**Figure 1**) [67, 68]. Soluble Mincle has been described to interact with *C. albicans* [68]. Moreover, in response to *C. albicans* infection, Mincle is required for TNF-α production from the macrophages [67]. Mincle also detects *Malassezia spp.* (human commensal fungi) and activates NFAT mediated cytokine transcription in cell lines. However, an impaired cytokine production and leukocyte recruitment was observed in Clec4e-/- macrophages in response to *Malassezia spp.* [69]. Currently, two glycolipids ligands have been identified from *C. albicans* that bind with Mincle. One is a polar glycolipid (comprising one dimannosyl-10-hydroxy-octadecanoic acid and two mannosyl-10-hydroxy-octadecanoic acids). The second ligand is a glyceroglycolipid containing the disaccharide gentiobiose joined to a glycerol backbone acylated with C14 and C18 fatty acids [51]. Although, Mincle ligands for other species of fungi have not been identified yet. However, it seems that Mincle must be specific in binding and recognizing glycolipid molecules among other pathogenic fungi [67, 68].

2.2.5 DC-sign

A type II transmembrane receptor with extracellular domain containing one CRD in its COOH- terminal and an and extracellular stalk. The stalk is comprised of seven residues that aid in DC-SIGN oligomerization [70, 71]. The CRD of DC-SIGN contains an EPF motif. DC-SIGN binds to glycoconjugates containing mannans and fucosylated carbohydrates in the presence of Ca^{2+} [72]. Both macrophages and dendritic cells express DC-SIGN. It is an endocytic receptor that can recognize and internalize several fungal pathogens followed by their release in the endosomal vesicles [73, 74]. In response to mannose-containing ligands, DC-SIGN has been described to enhance the cytokine production induced by the TLRs whereas fucosylated ligands amplify the IL-10 while inhibiting the production of proinflammatory cytokines. Mannosylated lipoarabinomannans (ManLAM) mediated activation of DC-SIGN results in inhibition of dendritic cells maturation by the LPS [75]. Thus, some pathogens infecting the dendritic cells can escape the immune activity by activating and inhibiting the DC-SIGN mediated maturation of dendritic cells. Thus, activation of DC-SIGN seems to exhibit complex effects such as internalizing the pathogens, triggering cytokine production and restricting the maturation of dendritic cells [76].

DC-SIGN has been involved in the recognition of many fungal pathogens including *C. albicans* (**Table 1**) (**Figure 1**). DC-SIGN is involved in the internalization and delivery of *C. albicans* to the phagolysosome of dendritic cells [77]. The binding and internalization of *C. albicans* is associated with the presence of N-linked mannans as a decreased binding was observed in *C. albicans* strains deficient in N-linked mannansylation [38]. However, dendritic cells do not exhibit any decreased binding affinity to the *C. albicans* strains lacking N-linked mannansylation. It appears that the binding of *C. albicans* through DC-SIGN in dendritic cells requires N-linked mannans whereas other glycoconjugates such as phosphomannans, O-linked mannans, or terminal β (1,2) mannosides are dispensable. There is no doubt regarding the role of DC-SIGN in recognition of *C. alibican* but it's the MR that in cooperation with DC-SIGN, contributes majorly to *C. albicans* internalization by the dendritic cells [38]. DC-SIGN is also an important PRR for recognition of *A. fumigatus* conidia by macrophages and dendritic cells. Unlike the *C. albicans*, MR is not required for the recognition of *A. fumigatus* by the human dendritic cells, however, this recognition by DC-SIGN has been described to be inhibited by the purified mannans and galacto-mannans [78]. Soluble DC-SIGN detects the glycoconjugates such as mannans, monosacchrides comprising mannans and Lewis antigen structures in a Ca^{2+} dependent manner [72, 74, 79]. Although, DC-SIGN is involved in the recognition of fungal pathogens, the underlying phenomena of modulation of macrophage and dendritic cell-mediated immune responses by DC-SIGN in response to fungi are still undefined.

2.3 CD11b/CD18 (MAC-1, CR3)

CD11b/CD18 also recognized as CD18 is a heterodimer receptor comprising of type I protein chains; αM chain (CD11b) and the common chain CD18 both of which are attached non-covalently. CD11b/CD18 is expressed by many of the leukocytes such as neutrophils, eosinophils, monocytes, macrophages and NK cells [80]. CD11b/CD18 helps in the adhesion of leukocytes to the activated endothelium and phagocytic receptors for antigens opsonized with iC3b [81]. In addition, CD11b/CD18 is also involved in the detection of β (1,3)-glucans. The αM chain of CD11b/CD18 possesses two distinct domains; the I-domain and a lectin domain. The I-domain binds ICAM-1, iC3b and fibrinogen whereas the lectin domain recognizes the fungal glycoconjugates such as β (1,3)-glucans, glucose, mannose and N-acetyl-D-glucosamine [82]. CD11b/CD18 triggers ROS production from neutrophils and macrophages in response to *S. cerevisiae* and zymosans [83]. Although, CD11b/CD18 is actively involved in the recognition of β (1,3)-glucans, however, some controversy exists in some experimental settings regarding its role in the identification of β (1,3)-glucans along with Dectin-1 mediated responses. The differences in experimental settings could be attributed to the observed disparities in results and can be attributed to many factors such as the use of distinct ligands (i.e., soluble vs. particulate β-glucan structures) [63, 84, 85], heterogeneity of β-glucan structures (both zymosan and fungi are heterogeneous and also contains carbohydrates and lipids in addition to mannans) [83], presence of the serum [85] and variability in the cell populations (neutrophils vs. macrophages) [84] used in the experiments. In conclusion, we can say that both CD11b/CD18 and Dectin-1 are involved in the recognition of β-glucan structures and their activation must lead to the induction of immune responses towards the fungal pathogens. Besides recognizing β-glucans, CD11b/CD18 also acts as an internalization receptor for mycobacterial PIM2 (a mycobacterial glycoconjugate-coated beads) [86] suggesting that CD11b/CD18 also work as a receptor for other fungal molecules other than β-glucan.

2.4 CD14

A glycosylphosphatidylinositol-anchored protein receptor was initially considered as an LPS binder. The Cd14 receptor is comprised of an extracellular domain containing cysteine-rich residues that form a horseshoe-like conformation [87–89]. Although, the CD14 receptor does not contain intracellular regions, however in cooperation with TLR2/MD receptors, it confers a high degree of sensitivity towards LPS [89]. CD14 has also been recognized as a co-receptor involved in TLR2- [90], TLR3- [91], TLR7- and TLR9-mediated detection of ligands [92]. Similar to the LPS, detection of mannans derived from *C. albicans* and *S. cerevisiae* also depends on CD14, LBP and TLR4 [14]. CD14 can also detect other fungal glycoconjugates, i.e., β (1,3)-glucans [21] and carbohydrate and therefore, act as an important receptor for innate immune recognition of *P. boydii* [93] and *A. fumiatus* [93]. However, the structural basis for the recognition of carbohydrate ligands by CD14 is still undefined. Also, the direct binding of CD14 with mannans and α-glucans has not been elucidated. There is a possibility that CD14 must be binding to these ligands *via* hydrophilic cleft. As CD14 also acts as a receptor for TLR ligands, we can speculate that CD14 must be promoting intracellular signaling first by binding to these carbohydrate ligands followed by their loading onto TLR2 or TLR4.

3. Fungal strategies of host innate immune evasion

3.1 Shielding of stimulatory PAMPs

Protecting the pathogen's inflammatory PAMPs from recognition by the host's PRRs is one of the most significant escape mechanisms employed by the microbes

[94, 95]. PRRs, which are found in various cellular components of primitive immune cells, are capable to identify recurrent pathogenic structures called PAMPs [96]. The host usually responds *via* phagocytic processes to establish the immediate antifungal mechanisms in response to fungal PAMPs. It is also accompanied by antimicrobial and pro-inflammatory responses launched through the activation of various intracellular signaling pathways that lead to the cytokine's and chemokine's gene transcription [97]. The prime objective of this response is to limit the disease while capturing as well as presenting the antigen to activate the adaptive immunity [98, 99]. NOD-like receptors (NLRs), TLRs, CTLRs and RIG-I-like, are the four groups of PRRs that vary in regards to ligand identification, signal transduction, as well as subcellular localization. Dendritic cells (DCs) and other myeloid cells exhibit the majority of PRRs, which are known for activating innate immune responses. PRR signaling, on the other hand, may regulate the progression of innate immune responses by the secretion of cytokines that helps in the polarization of $CD4^{+}$ cells [100]. CLRs are the main class of receptors that identify fungus, according to multiple investigations, whereas NRLs and TRLs play leading functions. Microbes may hide such that they are often overlooked by the immune system [101]. Polysaccharides as well as many other components of the cell wall are often layered and serve physiological and architectural roles in the cell wall. The structure of the cell wall layers of fungus is critical in serological identification [102]. Cell wall elements (i.e., chitin, mannan, and glucans) are also included among the fungal PAMPs. Most fungi contain chitin and also α (1,3)-glucan-based internal skeletal layer of the cell wall, which is linked to certain cell wall glycoproteins and polysaccharides [102]. Many fungal species may alter glucans and chitin to decrease host recognition, thus avoiding immune activation. Antagonism or synergism of receptor activation may result in a variety of diverse pathways of inflammatory processes. *In vivo*, a variety of fungal ligands have been exhibited in varying proportions, resulting in the activation of various PRRs [103]. Dectin-1 has a specific key for detecting hyphal infections *via* the identification of β-glucans, a fundamental component of the hyphal cell wall [94, 95]. *C. albicans* (a polymorphic fungus) may exhibit a transition between yeast and filamentous types, depending on environmental conditions. The bud scars on the *Candida* wall, which are exposed during budding, reveal the usually hidden β-glucan, and are predisposed to Dectin-1 identification. Usually, β-glucans of *C. albicans* are hidden from identification *via* Dectin-1 and outer wall elements during hyphal development [104, 105]. Dectin-1 also enhances the fungicidal activities of human neutrophils, which seem to be key effector cells throughout the fight against hyphal morphology [106]. The failure of Dectin-1 to identify the fungi as a result of β-glucan protection in hyphal development shields these bigger morphologies by avoiding internalization. In addition, hyphal types elicit protective T-helper cell type 2 (Th2) immunological responses in DCs rather than a Th1 immune response [107, 108]. Consequently, immune cells respond differentially to hyphae and yeast.

Similar to Dectin-1, Dectin-2 and Dectin-3 are also integral membrane proteins that belong to the CLRs family. Dectin-1 detects glucans, while Dectin-2/3 identifies mannans [109]. These may produce heterodimer complexes that provide greater sensitivity to host tissues as well as a high potential for binding to mannans [50]. While investigating the functions of Dectin-2 in *C. albicans* related-diseases, it was observed that the mice lacking Dectin-2 were more vulnerable to infection. In addition, phagocytosis and cytokine production was also decreased in these mice [110].

The α (1,3)-glucans present in the outermost layer of the cell wall helps in the pathogenicity of *Histoplasma capsulatum* (*H. capsulatum*) by hiding its immune-stimulatory β-glucans. This is analogous to how external mannans protect β-glucan from Dectin-1 recognition in *Candida*. Further evidence was observed in *H. Capsulatum* variants without α (1,3)-glucans which resulted in an increased TNF-α production. On the other hand, a reduction in Dectin-1 (necessary for β-glucans) expression, suppresses TNF-α levels. On switching into its infectious yeast stage, *Paracoccidioides brasiliensis* (*P. brasiliensis*) changes its β(1, 3)-glucans to α(1, 3)-glucans [111]. This is due to the reason that α (1, 3)-glucans are less likely to be identified by the host PRRs and therefore essential in fungal evasion of the immune system. *C. neoformans* masks the surface of PAMPs by producing GXM, which inhibits the production of IL-1β and pro-inflammatory TNF-α [112]. Both the pigment DHN-melanin and the RodA protein create a hydrophobic coating on *A. fumigatus* conidia and cover the glucans in order to avoid TLR activation. Although, quiescent conidia do not cause macrophages to secrete cytokines during germination, and the surface of RodA is destroyed. Furthermore, proteins that are recognized by PRRs are exposed to dendritic cells and macrophages, promoting the expression of the co-stimulatory molecule and cytokine production [113]. Dectin-1 redundant function may be explained by the lack of numerous glucans on the surface of resting *Aspergillus* conidia. Dectin-1 suppression on alveolar macrophages has little effect on the phagocytosis of *Aspergillus* that may be influenced by conidia germination [114, 115]. The spherule external wall glycoprotein (SOWgp) of *Coccidioides posadasii* (a respiratory fungal pathogen) is involved in its escape from innate immune recognition. The fungal cells secrete a metalloproteinase (Mep1) during endospore development, which metabolizes SOWgp [116]. Because SOWgp is downregulated during endospore production, fungal cells are therefore capable of avoiding phagocytosis and death during the susceptible spore-forming stage [117]. The capacity of the fungi to live in various morphotypes and to transiently shift from one form to another during infection is one strategy of shielding the host's immune system [118, 119]. These polymorphic phases are linked to phenotypic switching and result in evasion from cellular PRRs. This ability has most likely developed to help fungi survive in a variety of environments.

The *C. neoformans* capsule obscures the α (1,3)-glucans and mannan of the basal cell wall. Macrophages can easily recognize and phagocytose the acapsular mutant strains of *C. neoformans* and both glucans and mannose receptors are involved in this identification [120]. However, the capsule acts as a shield and masks the recognition of fungal PAMPs from the phagocytic receptor. Generally, TLRs detect the capsule and initiate an inflammatory response that is essential for limiting fungal infections. Further evidence in this regard has been provided by TLR2-deficient mice that have shown increased susceptibility to *C. neoformans* infections [121]. Contrary to yeast-form, blastoconidia of *C. albicans*, as well as hyphal form, are capable of evading the innate immune recognition by the Dectin-1 [104]. These blastoconidia activate the TLR2 and TLR4 in the ancillary monocytes and peritoneal macrophages. Such hyphal forms are not detected by the TLR4 and cause tissue-invasive infection [122]. Phenotypic change during germination could be a crucial survival strategy for several fungal pathogens. TLR4-mediated responses are diminished after the germination of *A. fumigatus* while its conidia elicit TLR2- as well as TLR4-mediated responses, in the tissue invasion. Mostly conidia germinate specifically into hyphae when TLR4 production is reduced, resulting in less intense proinflammatory cytokine production [123]. Proinflammatory responses mediated by TLR4 are essential in the prevention of invasive infections (aspergillosis) [124]. The stealth mode is not always successful and a fungal pathogen will usually be detected by the host in a certain way. Therefore, microorganisms frequently discover new strategies to exploit the host recognition networks and manipulate them for establishing a successful infection. These organisms may have chemicals on their surfaces or release compounds that trigger regulatory systems specifically. Throughout this way, the pathogen may either directly suppress or develop kinds of immune responses that aren't typically efficient against the pathogen [101].

3.2 Modulation of inflammatory signals

In respect of anti-inflammatory cytokine impact, TLR2 stimulation differentiates from TLR4 activation, with proinflammatory cytokine production being lower following TLR2 stimulation than the TLR4 stimulation [125]. TLR4 agonists selectively produced Th1-inducing cytokine signals in DCs, whereas TLR2 activation generate a more strong anti-inflammatory Th2 reaction [126]. Because each effector's arm elicits a different immune reaction, the equilibrium between Th1/Th2 reactions is thought to be important in deciding the severity of infection [127, 128]. The Th1 pathway generates pro-inflammatory cytokines such as IFN-γ, which stimulate cell-mediated immune mechanisms such as cytotoxicity and phagocyte activation. Th1 pathway is essential in the fight against intracellular and fungal infections. The Th2 pathway is characterized by cytokines such as IL-4, IL-5, as well as IL-10 and promotes a humoral response while suppressing the Th1-dependent effector functions [129]. Th2 cytokines may decrease monocyte anti-hyphal activity as well as lead to oxidative burst amid antifungal reactions [128]. TLR2-deficient macrophages have improved anti-candidal abilities [130], and TLR2 macrophages in mice are significantly more tolerant to widespread *C. albicans* related diseases [124, 130, 131]. As a result, the hyphal forms of *C. albicans* (tissue-invasive) and *A. fumigatus* are likely to shift the equilibrium towards that Th2 pathway by avoiding TLR4 stimulation in favor of TLR2 activation. *C. neoformans* has also been found to possess immunosuppressive properties. The primary virulence component of *C. neoformans* is indeed the GXM, which is a strong activator of the IL-10 (anti-inflammatory cytokine) and a pro-Th2 cytokine mediator in human monocytes [132, 133]. The melanin pigment produced frequently by pathogenic filamentous fungi has been associated with fungal pathogenicity and its immunomodulatory impact in *C. neoformans* has been investigated. Melanized *C. neoformans* variants result in increased pulmonary IL-4 levels thus driving the host cells to switch towards Th2 response [134]. *Blastomyces dermatitidis* (*B. dermatitidis*) that causes systemic and pulmonary mycosis, may result in comparative immunosuppression by reducing the synthesis of the TNF-α (a pro-inflammatory cytokine) [135]. The binding of Blastomyces surface adhesins to the complement receptor III on macrophages results in suppression of TNF-α synthesis which otherwise could be harmful to *B. dermatitidis*' existence [136].

3.3 Shedding of decoy components

Several innate immune evasion strategies have been identified for *Pneumocystis jiroveci* (*P. jiroveci*), an opportunistic fungi that usually infects AIDS patients. Glycoprotein A (gpA) complex is the main protein antigen present on the membrane of *Pneumocystis*. It is highly glycosylated with glucose, mannose, as well as galactose-containing carbohydrate moieties [137]. Mannose receptors present on the alveolar monocytes recognize these structures. However, *Pneumocystis* escapes this recognition by MR on alveolar macrophages and impedes its phagocytic activity by premature release of its gpA glycoprotein as a decoy [138]. Furthermore, *Pneumocystis* has also been described to deplete MR from the membrane of AMs, preventing non-opsonic absorption by MR (surface-expressed) [139].

3.4 Persistence in the intracellular environments

Several fungal pathogens have developed the potential to avoid the phagocytic activity of macrophages. For example, *C. neoformans* can phenotypically shift to a mucilaginous colony type generating a significantly bigger capsulated polysaccharide GXM with modified biochemical and biophysical characteristics thus limiting AM phagocytic effectiveness [140]. When certain variants fail to escape host identification, phagocytosis by macrophages does not necessarily result in death and the ending of the life cycle as some fungi may survive the harsh environment inside the phagolysosome. Some *C. albicans spp.* can withstand intracellular death and produce hyphal structures and thus escape the macrophages [141]. *C. albicans* have an extremely specialized anti-nitric oxide (NO) defense mechanism including the NO-scavenging flavohemoglobin genetic traits that convert NO to less toxic substances when comes into contact with reactive nitrogen molecules like NO and oxygen free radicals generated by monocytes/macrophages [142]. Comprehensive morphologic investigations have shown that *Candida* could produce germ tubes, proliferate, and ultimately escape the host cell despite phagocytosis by macrophages [143]. Phagocytosis provokes *C. albicans* within macrophages to switch into self-preservation mode, which includes a delayed growth rate, carbon utilization, as well as an oxidative stress reaction to thrive in the hostile environment inside macrophages [144, 145] suggesting that the phagocytic activity solely might not be sufficient to clear the infection from the host. During persistent infection, the fungi have been shown to survive and reproduce inside the phagocytic cells [146, 147]. To escape intracellular death, *C. neoformans* cause aberrant lysosomal transport and significant cytoplasmic vacuolation in the host cell, leading to host cell disintegration [148]. Similarly, *H. capsulatum*, is also capable of surviving inside macrophages for longer durations following primary infection and become activated as the immune responses are diminished [149, 150]. *Histoplasma* is supposed to prevent phagolysosome formation and proactively regulate the phagosomal pH following phagocytosis to maximize its survival inside phagosomes [151, 152]. Furthermore, *Histoplasma* can prevent the production of toxic superoxide radicals that are harmful to its survival inside macrophages [153].

3.5 Complement evasion

The complement system is a dynamic mechanism that plays a significant part in innate immunity and antibody-mediated protection against pathogenic microbes [154]. Several foreign antigens including fungal PAMPs, cellular debris, as well as antigen–antibody complexes can activate a series of complement pathways [98, 155]. Excessive tissue damage and inflammation by the complement system are avoided by the regulatory molecules of the complement system [156]. The complement system is split into three pathways; classical, alternative and lectin pathway. The activation of all these pathways varies in regards to associated components but all pathways submerge by producing the same group of effector molecules, i.e., opsonization and formation of membrane attack complex (MAC) [96]. All complement mechanisms contribute to the production of C3 convertase as well as the C3b fraction, which in turn promotes the synthesis of C5 convertase. C5 convertase cleaves the C5 factor into C5a and C5b. The distal complement components are formed as a result of a succession of accumulation and polymerization processes, as well as the mobilization of terminal complement elements such as C6, C7, C8, and C9. The terminal complement components form MAC causing cell lysis by inserting C9 into the lipid membrane layer [157–159]. Pathogenic organisms, on the other hand, have adopted different approaches to evade complement attacks, such as binding to regulatory complement proteins by secreting proteases or evading opsonization. For example, *Aspergillus spp.* and *C. albicans* release proteins on their membranes that bind to complement proteins to prevent being eliminated by the complement system. These proteins, when linked to the hyphal surface, block the complement cascade, allowing the fungi to avoid the complement attack [156]. *Aspergillus* and *Candida* are recognized to activate complement by depositing C3 on the fungal membrane, which facilitates opsonization and the synthesis of the chemoattractant (C5a), which recruits leukocytes to the infected area [160–162]. Pigmentation on the conidial surface of *A. fumigatus* has been demonstrated to influence pathogenicity by reducing C3 protein accumulation and neutrophil activity [163]. Transcription factors such as Factor H-like protein 1 (FHL1), Factor H, as well as C4 binding protein (C4BP) for the signaling pathway, keep the complement system in balance against abnormal activation. *A. fumigatus* and *C. albicans* both have been described to bind FHL-1, C4BP, and Factor H, on their membrane to evade the complement cascade [164–166]. Furthermore, the dense yeast cell wall is impervious to immediate lysis by the MAC [161]. Complement is far more than a "defensive" mechanism against infections. It has a role in inflammatory responses, cellular response regulation, and cell–cell interactions, all of which are important for cell differentiation and initial growth [167]. Currently, two complement-targeted drugs for non-fungal illnesses have been approved in the health center: eculizumab (an anti-C5 antibody) and different formulations of C1 esterase inhibitor (C1-INH). Several other drugs that target distinct elements of the complement cascade are all in different phases of trials [167–170].

4. Conclusion

Our understanding regarding the innate immune recognition of pathogenic fungi by the corresponding fungal PAMPs is still poor. Moreover, the fungal ligands involved in the activation of host PRRs remain largely unknown for several pathogenic fungi. Characterization of fungal PAMPs and their recognition by the host PRRs can provide a comprehensive understanding of pathogenesis and immunity to the pathogenic fungi. In addition, characterization of these fungal ligands and their activation of respective PRRs is essential not only to discover new therapeutic approaches against fungal infections particularly in immune-compromised patients but also to develop novel adjuvants for enhancing the prophylactic immune responses against pathogenic fungi.

Acknowledgements

The author would like to acknowledge all the co-authors for their contribution.

Conflict of interest

No competing financial interest exists.

Author details

Faisal Rasheed Anjum[1,2*,†], Sidra Anam[2*,†], Muhammad Luqman[2], Ameena A. AL-surhanee[3], Abdullah F. Shater[4], Muhammad Wasim Usmani[5], Sajjad ur Rahman[2], Muhammad Sohail Sajid[1,6], Farzana Rizvi[5] and Muhammad Zulqarnain Shakir[5]

1 Department of Epidemiology and Public Health, University of Agriculture, Faisalabad, Pakistan

2 Institute of Microbiology, University of Agriculture Faisalabad, Pakistan

3 Biology Department, College of Science, Jouf University, Sakaka, Kingdom of Saudi Arabia

4 Faculty of Applied Medical Sciences, Department of Medical Laboratory Technology, University of Tabuk, Tabuk, Kingdom of Saudi Arabia

5 Department of Pathology, Faculty of Veterinary Science, University of Agriculture, Faisalabad, Pakistan

6 Department of Parasitology, Faculty of Veterinary Science, University of Agriculture, Faisalabad, Pakistan

*Address all correspondence to: drfaissalatarar@gmail.com and sidraanam2924@gmail.com

† Both authors contributed equally and the order was determined by coin flip.

References

[1] Singh N. Trends in the epidemiology of opportunistic fungal infections: predisposing factors and the impact of antimicrobial use practices. Clinical Infectious Diseases. 2001;**33**(10): 1692-1696

[2] Segal BH, Walsh TJ. Current approaches to diagnosis and treatment of invasive aspergillosis. American Journal of Respiratory and Critical Care Medicine. 2006;**173**(7):707-717

[3] Morgan J et al. Incidence of invasive aspergillosis following hematopoietic stem cell and solid organ transplantation: interim results of a prospective multicenter surveillance program. Medical Mycology. 2005;**43**(Supplement_1):S49-S58

[4] Chamilos G et al. Invasive fungal infections in patients with hematologic malignancies in a tertiary care cancer center: an autopsy study over a 15-year period (1989-2003). Haematologica. 2006;**91**(7):986-989

[5] Marr KA et al. Epidemiology and outcome of mould infections in hematopoietic stem cell transplant recipients. Clinical Infectious Diseases. 2002;**34**(7):909-917

[6] Pasqualotto A, Denning D. Post-operative aspergillosis. Clinical Microbiology and Infection. 2006;**12**(11):1060-1076

[7] Meersseman W et al. Galactomannan in bronchoalveolar lavage fluid: a tool for diagnosing aspergillosis in intensive care unit patients. American Journal of Respiratory and Critical Care Medicine. 2008;**177**(1):27-34

[8] Akira S, Uematsu S, Takeuchi O. Pathogen recognition and innate immunity. Cell. 2006;**124**(4):783-801

[9] Medzhitov R. Recognition of microorganisms and activation of the immune response. Nature. 2007;**449**(7164):819-826

[10] van de Veerdonk FL et al. Host–microbe interactions: Innate pattern recognition of fungal pathogens. Current Opinion in Microbiology. 2008;**11**(4):305-312

[11] Takeuchi O, Akira S. Pattern recognition receptors and inflammation. Cell. 2010;**140**(6):805-820

[12] Hidmark A, von Saint Paul A, Dalpke AH. Cutting edge: TLR13 is a receptor for bacterial RNA. The Journal of Immunology. 2012;**189**(6):2717-2721

[13] Oldenburg M et al. TLR13 recognizes bacterial 23S rRNA devoid of erythromycin resistance–forming modification. Science. 2012;**337**(6098): 1111-1115

[14] Tada H et al. Saccharomyces cerevisiae-and Candida albicans-derived mannan induced production of tumor necrosis factor alpha by human monocytes in a CD14-and Toll-like receptor 4-dependent manner. Microbiology and Immunology. 2002;**46**(7):503-512

[15] Netea MG et al. Immune sensing of Candida albicans requires cooperative recognition of mannans and glucans by lectin and Toll-like receptors. The Journal of Clinical Investigation. 2006;**116**(6):1642-1650

[16] Figueiredo RT et al. TLR4 recognizes Pseudallescheria boydii conidia and purified rhamnomannans. Journal of Biological Chemistry. 2010;**285**(52): 40714-40723

[17] Shoham S et al. Toll-like receptor 4 mediates intracellular signaling without TNF-α release in response to Cryptococcus neoformans polysaccharide capsule. The Journal of Immunology. 2001;**166**(7):4620-4626

[18] Ibata-Ombetta S et al. Candida albicans phospholipomannan promotes survival of phagocytosed yeasts through modulation of bad phosphorylation and macrophage apoptosis. Journal of Biological Chemistry. 2003;**278**(15): 13086-13093

[19] Kakutani R et al. Essential role of Toll-like receptor 2 in macrophage activation by glycogen. Glycobiology. 2012;**22**(1):146-159

[20] Fonseca FL et al. Immunomodulatory effects of serotype B glucuronoxylomannan from Cryptococcus gattii correlate with polysaccharide diameter. Infection and Immunity. 2010;**78**(9):3861-3870

[21] Bittencourt VCB et al. An α-glucan of Pseudallescheria boydii is involved in fungal phagocytosis and toll-like receptor activation. Journal of Biological Chemistry. 2006;**281**(32):22614-22623

[22] Hsu T-L et al. Profiling carbohydrate-receptor interaction with recombinant innate immunity receptor-Fc fusion proteins. Journal of Biological Chemistry. 2009;**284**(50): 34479-34489

[23] Jin MS et al. Crystal structure of the TLR1-TLR2 heterodimer induced by binding of a tri-acylated lipopeptide. Cell. 2007;**130**(6):1071-1082

[24] Kang JY et al. Recognition of lipopeptide patterns by Toll-like receptor 2-Toll-like receptor 6 heterodimer. Immunity. 2009;**31**(6):873-884

[25] Park BS et al. The structural basis of lipopolysaccharide recognition by the TLR4–MD-2 complex. Nature. 2009; **458**(7242):1191-1195

[26] Figueiredo RT, Carneiro LA, Bozza MT. Fungal surface and innate immune recognition of filamentous fungi. Frontiers in Microbiology. 2011;**2**:248

[27] Drickamer K. Engineering galactose-binding activity into a C-type mannose-binding protein. Nature. 1992;**360**(6400):183-186

[28] Kolatkar AR, Weis WI. Structural basis of galactose recognition by C-type animal lectins (*). Journal of Biological Chemistry. 1996;**271**(12):6679-6685

[29] Kolatkar AR et al. Mechanism of N-acetylgalactosamine binding to a C-type animal lectin carbohydrate-recognition domain. Journal of Biological Chemistry. 1998;**273**(31): 19502-19508

[30] Zelensky AN, Gready JE. The C-type lectin-like domain superfamily. The FEBS Journal. 2005;**272**(24): 6179-6217

[31] Sancho D, Sousa CR e. Signaling by myeloid C-type lectin receptors in immunity and homeostasis. Annual Review of Immunology. 2012;**30**: 491-529

[32] Martinez-Pomares L. The mannose receptor. Journal of Leukocyte Biology. 2012;**92**(6):1177-1186

[33] Taylor ME, Drickamer K. Structural requirements for high affinity binding of complex ligands by the macrophage mannose receptor. Journal of Biological Chemistry. 1993;**268**(1):399-404

[34] Taylor ME, Bezouska K, Drickamer K. Contribution to ligand binding by multiple carbohydrate-recognition domains in the macrophage mannose receptor. Journal of Biological Chemistry. 1992;**267**(3):1719-1726

[35] Liu Y et al. Crystal structure of the cysteine-rich domain of mannose receptor complexed with a sulfated carbohydrate ligand. Journal of Experimental Medicine. 2000;**191**(7):1105-1116

[36] Leteux C et al. The cysteine-rich domain of the macrophage mannose

receptor is a multispecific lectin that recognizes chondroitin sulfates A and B and sulfated oligosaccharides of blood group Lewisa and Lewisx types in addition to the sulfated N-glycans of lutropin. Journal of Experimental Medicine. 2000;**191**(7):1117-1126

[37] Ezekowitz R et al. Uptake of Pneumocystis carinii mediated by the macrophage mannose receptor. Nature. 1991;**351**(6322):155-158

[38] Cambi A et al. Dendritic cell interaction with Candida albicans critically depends on N-linked mannan. Journal of Biological Chemistry. 2008;**283**(29):20590-20599

[39] Le Cabec V et al. The human macrophage mannose receptor is not a professional phagocytic receptor. Journal of Leukocyte Biology. 2005;**77**(6):934-943

[40] Tachado SD et al. Pneumocystis-mediated IL-8 release by macrophages requires coexpression of mannose receptors and TLR2. Journal of Leukocyte Biology. 2007;**81**(1):205-211

[41] van de Veerdonk FL et al. The macrophage mannose receptor induces IL-17 in response to Candida albicans. Cell Host & Microbe. 2009;**5**(4):329-340

[42] Kang PB et al. The human macrophage mannose receptor directs Mycobacterium tuberculosis lipoarabinomannan-mediated phagosome biogenesis. The Journal of Experimental Medicine. 2005;**202**(7): 987-999

[43] Torrelles JB, Azad AK, Schlesinger LS. Fine discrimination in the recognition of individual species of phosphatidyl-myo-inositol mannosides from Mycobacterium tuberculosis by C-type lectin pattern recognition receptors. The Journal of Immunology. 2006;**177**(3):1805-1816

[44] Sato K et al. Dectin-2 is a pattern recognition receptor for fungi that couples with the Fc receptor γ chain to induce innate immune responses. Journal of Biological Chemistry. 2006;**281**(50):38854-38866

[45] Barrett NA et al. Dectin-2 recognition of house dust mite triggers cysteinyl leukotriene generation by dendritic cells. The Journal of Immunology. 2009;**182**(2):1119-1128

[46] Saijo S et al. Dectin-2 recognition of α-mannans and induction of Th17 cell differentiation is essential for host defense against Candida albicans. Immunity. 2010;**32**(5):681-691

[47] Ariizumi K et al. Cloning of a second dendritic cell-associated C-type lectin (dectin-2) and its alternatively spliced isoforms. Journal of Biological Chemistry. 2000;**275**(16):11957-11963

[48] Robinson MJ et al. Dectin-2 is a Syk-coupled pattern recognition receptor crucial for Th17 responses to fungal infection. Journal of Experimental Medicine. 2009;**206**(9): 2037-2051

[49] McGreal EP et al. The carbohydrate-recognition domain of Dectin-2 is a C-type lectin with specificity for high mannose. Glycobiology. 2006;**16**(5): 422-430

[50] Zhu L-L et al. C-type lectin receptors Dectin-3 and Dectin-2 form a heterodimeric pattern-recognition receptor for host defense against fungal infection. Immunity. 2013;**39**(2): 324-334

[51] Ishikawa T et al. Identification of distinct ligands for the C-type lectin receptors Mincle and Dectin-2 in the pathogenic fungus Malassezia. Cell Host & Microbe. 2013;**13**(4):477-488

[52] Brown GD et al. Dectin-1 is a major β-glucan receptor on macrophages. The

Journal of Experimental Medicine. 2002;**196**(3):407-412

[53] Taylor PR et al. The β-glucan receptor, dectin-1, is predominantly expressed on the surface of cells of the monocyte/macrophage and neutrophil lineages. The Journal of Immunology. 2002;**169**(7):3876-3882

[54] Willment JA et al. The human β-glucan receptor is widely expressed and functionally equivalent to murine Dectin-1 on primary cells. European Journal of Immunology. 2005;**35**(5): 1539-1547

[55] Brown GD et al. Dectin-1 mediates the biological effects of β-glucans. The Journal of Experimental Medicine. 2003;**197**(9):1119-1124

[56] Yoshitomi H et al. A role for fungal β-glucans and their receptor Dectin-1 in the induction of autoimmune arthritis in genetically susceptible mice. Journal of Experimental Medicine. 2005;**201**(6): 949-960

[57] Gantner BN et al. Collaborative induction of inflammatory responses by dectin-1 and Toll-like receptor 2. Journal of Experimental Medicine. 2003;**197**(9): 1107-1117

[58] Adams EL et al. Differential high-affinity interaction of dectin-1 with natural or synthetic glucans is dependent upon primary structure and is influenced by polymer chain length and side-chain branching. Journal of Pharmacology and Experimental Therapeutics. 2008;**325**(1):115-123

[59] Brown GD, Gordon S. A new receptor for β-glucans. Nature. 2001; **413**(6851):36-37

[60] Rogers NC et al. Syk-dependent cytokine induction by Dectin-1 reveals a novel pattern recognition pathway for C type lectins. Immunity. 2005;**22**(4):507-517

[61] Goodridge HS et al. Activation of the innate immune receptor Dectin-1 upon formation of a 'phagocytic synapse'. Nature. 2011;**472**(7344): 471-475

[62] Goodridge HS, Simmons RM, Underhill DM. Dectin-1 stimulation by Candida albicans yeast or zymosan triggers NFAT activation in macrophages and dendritic cells. The Journal of Immunology. 2007;**178**(5): 3107-3115

[63] Rosas M et al. The induction of inflammation by dectin-1 in vivo is dependent on myeloid cell programming and the progression of phagocytosis. The Journal of Immunology. 2008;**181**(5):3549-3557

[64] Goodridge HS et al. Differential use of CARD9 by dectin-1 in macrophages and dendritic cells. The Journal of Immunology. 2009;**182**(2):1146-1154

[65] Matsumoto M et al. A novel LPS-inducible C-type lectin is a transcriptional target of NF-IL6 in macrophages. The Journal of Immunology. 1999;**163**(9):5039-5048

[66] Yamasaki S et al. Mincle is an ITAM-coupled activating receptor that senses damaged cells. Nature Immunology. 2008;**9**(10):1179-1188

[67] Wells CA et al. The macrophage-inducible C-type lectin, mincle, is an essential component of the innate immune response to Candida albicans. The Journal of Immunology. 2008; **180**(11):7404-7413

[68] Bugarcic A et al. Human and mouse macrophage-inducible C-type lectin (Mincle) bind Candida albicans. Glycobiology. 2008;**18**(9):679-685

[69] Yamasaki S et al. C-type lectin Mincle is an activating receptor for pathogenic fungus, Malassezia. Proceedings of the National Academy of Sciences. 2009;**106**(6):1897-1902

[70] Geijtenbeek TB et al. Identification of DC-SIGN, a novel dendritic cell–specific ICAM-3 receptor that supports primary immune responses. Cell. 2000;**100**(5):575-585

[71] Mitchell DA, Fadden AJ, Drickamer K. A novel mechanism of carbohydrate recognition by the C-type lectins DC-SIGN and DC-SIGNR: Subunit organization and binding to multivalent ligands. Journal of Biological Chemistry. 2001;**276**(31): 28939-28945

[72] Appelmelk BJ et al. Cutting edge: carbohydrate profiling identifies new pathogens that interact with dendritic cell-specific ICAM-3-grabbing nonintegrin on dendritic cells. The Journal of Immunology. 2003;**170**(4): 1635-1639

[73] Kwon DS et al. DC-SIGN-mediated internalization of HIV is required for trans-enhancement of T cell infection. Immunity. 2002;**16**(1):135-144

[74] Guo Y et al. Structural basis for distinct ligand-binding and targeting properties of the receptors DC-SIGN and DC-SIGNR. Nature Structural & Molecular Biology. 2004;**11**(7):591-598

[75] Gringhuis SI et al. Carbohydrate-specific signaling through the DC-SIGN signalosome tailors immunity to Mycobacterium tuberculosis, HIV-1 and Helicobacter pylori. Nature Immunology. 2009;**10**(10):1081-1088

[76] van Kooyk Y, Geijtenbeek TB. DC-SIGN: Escape mechanism for pathogens. Nature Reviews Immunology. 2003;**3**(9):697-709

[77] Cambi A et al. The C-type lectin DC-SIGN (CD209) is an antigen-uptake receptor for Candida albicans on dendritic cells. European Journal of Immunology. 2003;**33**(2):532-538

[78] Serrano-Gómez D et al. Dendritic cell-specific intercellular adhesion molecule 3-grabbing nonintegrin mediates binding and internalization of Aspergillus fumigatus conidia by dendritic cells and macrophages. The Journal of Immunology. 2004;**173**(9): 5635-5643

[79] Van Liempt E et al. Molecular basis of the differences in binding properties of the highly related C-type lectins DC-SIGN and L-SIGN to Lewis X trisaccharide and Schistosoma mansoni egg antigens. Journal of Biological Chemistry. 2004;**279**(32):33161-33167

[80] Ross GD. Regulation of the adhesion versus cytotoxic functions of the Mac-1/CR3/α M β 2-lntegrin glycoprotein. Critical Reviews in Immunology. 2000;**20**(3):1-30

[81] Holers VM. Complement and its receptors: new insights into human disease. Annual Review of Immunology. 2014;**32**:433-459

[82] Thornton BP et al. Analysis of the sugar specificity and molecular location of the beta-glucan-binding lectin site of complement receptor type 3 (CD11b/CD18). The Journal of Immunology. 1996;**156**(3):1235-1246

[83] van Bruggen R et al. Complement receptor 3, not Dectin-1, is the major receptor on human neutrophils for β-glucan-bearing particles. Molecular Immunology. 2009;**47**(2-3):575-581

[84] Qi C et al. Differential pathways regulating innate and adaptive antitumor immune responses by particulate and soluble yeast-derived β-glucans. Blood, The Journal of the American Society of Hematology. 2011;**117**(25):6825-6836

[85] Bose N et al. Binding of soluble yeast β-glucan to human neutrophils and monocytes is complement-dependent. Frontiers in Immunology. 2013;**4**:230

[86] Villeneuve C et al. Mycobacteria use their surface-exposed glycolipids to infect human macrophages through a receptor-dependent process. Journal of Lipid Research. 2005;**46**(3):475-483

[87] Kim J-I et al. Crystal Structure of CD14 and Its Implications for Lipopolysaccharide Signaling*♦. Journal of Biological Chemistry. 2005;**280**(12):11347-11351

[88] Kelley SL et al. The crystal structure of human soluble CD14 reveals a bent solenoid with a hydrophobic amino-terminal pocket. The Journal of Immunology. 2013;**190**(3):1304-1311

[89] Granucci F, Zanoni I. Role of CD14 in host protection against infections and in metabolism regulation. Frontiers in Cellular and Infection Microbiology. 2013;**3**:32

[90] Schröder NW et al. Lipopolysaccharide binding protein binds to triacylated and diacylated lipopeptides and mediates innate immune responses. The Journal of Immunology. 2004;**173**(4):2683-2691

[91] Lee H-K et al. Double-stranded RNA-mediated TLR3 activation is enhanced by CD14. Immunity. 2006;**24**(2):153-163

[92] Baumann CL et al. CD14 is a coreceptor of Toll-like receptors 7 and 9CD14: Coreceptor of TLR7 and TLR9. The Journal of Experimental Medicine. 2010;**207**(12):2689-2701

[93] Wang J et al. Involvement of CD14 and toll-like receptors in signalling by Aspergillus hyphae. British Journal of Surgery. 2001;**88**(8):1138-1150

[94] Latgé JP et al. Specific molecular features in the organization and biosynthesis of the cell wall of Aspergillus fumigatus. Medical Mycology. 2005;**43**(Supplement_1): S15-S22

[95] Gow NA et al. Immune recognition of Candida albicans β-glucan by dectin-1. The Journal of Infectious Diseases. 2007;**196**(10):1565-1571

[96] Janeway C et al. Basic concepts in immunology. In: Immunobiology the Immune System in Health and Disease. 5th ed. New York: Garland Publishing; 2001. pp. 1-34

[97] Shizuo A, Takeda K. Toll-like receptor signaling. Nature Reviews. Immunology. 2004;**4**(7):499-511

[98] Chai LY et al. Fungal strategies for overcoming host innate immune response. Medical Mycology. 2009; **47**(3):227-236

[99] Bachiega TF et al. Participation of dectin-1 receptor on NETs release against Paracoccidioides brasiliensis: Role on extracellular killing. Immunobiology. 2016;**221**(2):228-235

[100] Plato A, Hardison SE, Brown GD. Pattern recognition receptors in antifungal immunity. Seminars in Immunopathology. Springer Berlin Heidelberg. 2015;**37**(2):97-106

[101] Underhill DM. Escape mechanisms from the immune response. In: Immunology of Fungal Infections. Dordrecht: Springer; 2007. pp. 429-442

[102] Erwig LP, Gow NA. Interactions of fungal pathogens with phagocytes. Nature Reviews Microbiology. 2016; **14**(3):163-176

[103] Levitz SM. Innate recognition of fungal cell walls. PLoS Pathogens. 2010;**6**(4):e1000758

[104] Gantner BN, Simmons RM, Underhill DM. Dectin-1 mediates macrophage recognition of Candida albicans yeast but not filaments. The EMBO Journal. 2005;**24**(6):1277-1286

[105] Heinsbroek SE, Brown GD, Gordon S. Dectin-1 escape by fungal

dimorphism. Trends in Immunology. 2005;**26**(7):352-354

[106] Kennedy AD et al. Dectin-1 promotes fungicidal activity of human neutrophils. European Journal of Immunology. 2007;**37**(2):467-478

[107] Romani L, Bistoni F, Puccetti P. Fungi, dendritic cells and receptors: a host perspective of fungal virulence. Trends in Microbiology. 2002;**10**(11): 508-514

[108] d'Ostiani CF et al. Dendritic cells discriminate between yeasts and hyphae of the fungus Candida albicansImplications for initiation of T helper cell immunity in vitro and in vivo. Journal of Experimental Medicine. 2000;**191**(10):1661-1674

[109] Saijo S, Iwakura Y. Dectin-1 and Dectin-2 in innate immunity against fungi. International Immunology. 2011;**23**(8):467-472

[110] Ifrim DC et al. The role of dectin-2 for host defense against disseminated candidiasis. Journal of Interferon & Cytokine Research. 2016;**36**(4):267-276

[111] Borges-Walmsley MI et al. The pathobiology of Paracoccidioides brasiliensis. Trends in Microbiology. 2002;**10**(2):80-87

[112] Vecchiarelli A et al. Downregulation by cryptococcal polysaccharide of tumor necrosis factor alpha and interleukin-1 beta secretion from human monocytes. Infection and Immunity. 1995;**63**(8):2919-2923

[113] Aimanianda V et al. Erratum: Surface hydrophobin prevents immune recognition of airborne fungal spores. Nature. 2010;**465**(7300):966-966

[114] Steele C et al. The beta-glucan receptor dectin-1 recognizes specific morphologies of Aspergillus fumigatus. PLoS Pathogens. 2005;**1**(4):e42

[115] Slesiona S et al. Persistence versus escape: Aspergillus terreus and Aspergillus fumigatus employ different strategies during interactions with macrophages. PLoS One. 2012;**7**(2): e31223

[116] Hung C-Y et al. A metalloproteinase of Coccidioides posadasii contributes to evasion of host detection. Infection and Immunity. 2005;**73**(10):6689-6703

[117] Hung C-Y et al. A parasitic phase-specific adhesin of Coccidioides immitis contributes to the virulence of this respiratory fungal pathogen. Infection and Immunity. 2002;**70**(7):3443-3456

[118] Romani L. Immunity to fungal infections. Nature Reviews Immunology. 2004;**4**(1):11-24

[119] Hogan LH, Klein BS, Levitz SM. Virulence factors of medically important fungi. Clinical Microbiology Reviews. 1996;**9**(4):469-488

[120] Cross C, Bancroft G. Ingestion of acapsular Cryptococcus neoformans occurs via mannose and beta-glucan receptors, resulting in cytokine production and increased phagocytosis of the encapsulated form. Infection and Immunity. 1995;**63**(7):2604-2611

[121] Yauch LE et al. Involvement of CD14, toll-like receptors 2 and 4, and MyD88 in the host response to the fungal pathogen Cryptococcus neoformans in vivo. Infection and Immunity. 2004;**72**(9):5373-5382

[122] van der Graaf CA et al. Differential cytokine production and Toll-like receptor signaling pathways by Candida albicans blastoconidia and hyphae. Infection and Immunity. 2005;**73**(11): 7458-7464

[123] Netea MG et al. Aspergillus fumigatus evades immune recognition during germination through loss of toll-like receptor-4-mediated signal

transduction. The Journal of Infectious Diseases. 2003;**188**(2):320-326

[124] Bellocchio S et al. The contribution of the Toll-like/IL-1 receptor superfamily to innate and adaptive immunity to fungal pathogens in vivo. The Journal of Immunology. 2004; **172**(5):3059-3069

[125] Hirschfeld M et al. Signaling by toll-like receptor 2 and 4 agonists results in differential gene expression in murine macrophages. Infection and Immunity. 2001;**69**(3):1477-1482

[126] Re F, Strominger JL. Toll-like receptor 2 (TLR2) and TLR4 differentially activate human dendritic cells. Journal of Biological Chemistry. 2001;**276**(40):37692-37699

[127] Netea MG et al. From the Th1/Th2 paradigm towards a Toll-like receptor/T-helper bias. Antimicrobial Agents and Chemotherapy. 2005;**49**(10):3991-3996

[128] Stevens DA. Th1/Th2 in aspergillosis. Medical Mycology. 2006;**44**(Supplement_1):S229-S235

[129] Jankovic D, Liu Z, Gause WC. Th1-and Th2-cell commitment during infectious disease: asymmetry in divergent pathways. Trends in Immunology. 2001;**22**(8):450-457

[130] Blasi E et al. Biological importance of the two Toll-like receptors, TLR2 and TLR4, in macrophage response to infection with Candida albicans. FEMS Immunology and Medical Microbiology. 2005;**44**(1):69-79

[131] Netea MG et al. Toll-like receptor 2 suppresses immunity against Candida albicans through induction of IL-10 and regulatory T cells. The Journal of Immunology. 2004;**172**(6):3712-3718

[132] Vecchiarelli A et al. Purified capsular polysaccharide of Cryptococcus neoformans induces interleukin-10 secretion by human monocytes. Infection and Immunity. 1996;**64**(7):2846-2849

[133] Chiapello LS et al. Immunosuppression, interleukin-10 synthesis and apoptosis are induced in rats inoculated with Cryptococcus neoformans glucuronoxylomannan. Immunology. 2004;**113**(3):392-400

[134] Mednick AJ, Nosanchuk JD, Casadevall A. Melanization of Cryptococcus neoformans affects lung inflammatory responses during cryptococcal infection. Infection and Immunity. 2005;**73**(4):2012-2019

[135] Finkel-Jimenez B et al. The WI-1 adhesin blocks phagocyte TNF-α production, imparting pathogenicity on Blastomyces dermatitidis. The Journal of Immunology. 2001;**166**(4):2665-2673

[136] Brandhorst TT et al. Exploiting type 3 complement receptor for TNF-α suppression, immune evasion, and progressive pulmonary fungal infection. The Journal of Immunology. 2004; **173**(12):7444-7453

[137] Stringer JR, Keely SP. Genetics of surface antigen expression in Pneumocystis carinii. Infection and Immunity. 2001;**69**(2):627-639

[138] Pop SM, Kolls JK, Steele C. Pneumocystis: Immune recognition and evasion. The International Journal of Biochemistry & Cell Biology. 2006;**38**(1):17-22

[139] Fraser IP et al. Pneumocystis carinii enhancessoluble mannose receptor production by macrophages. Microbes and Infection. 2000;**2**(11): 1305-1310

[140] Fries BC et al. Phenotypic switching of Cryptococcus neoformans occurs in vivo and influences the outcome of infection. The Journal of Clinical Investigation. 2001;**108**(11): 1639-1648

[141] Tavanti A et al. Candida albicans isolates with different genomic backgrounds display a differential response to macrophage infection. Microbes and Infection. 2006;**8**(3): 791-800

[142] Ullmann BD et al. Inducible defense mechanism against nitric oxide in Candida albicans. Eukaryotic Cell. 2004;**3**(3):715-723

[143] Káposzta R et al. Rapid recruitment of late endosomes and lysosomes in mouse macrophages ingesting Candida albicans. Journal of Cell Science. 1999;**112**(19):3237-3248

[144] Lorenz MC, Bender JA, Fink GR. Transcriptional response of Candida albicans upon internalization by macrophages. Eukaryotic Cell. 2004;**3**(5):1076-1087

[145] Richardson MD, Smith H. Resistance of virulent and attenuated strains of Candida albicans to Intracellular Killing by Human and Mouse Phagocytes. The Journal of Infectious Diseases. 1981;**144**(6): 557-564

[146] Goldman DL et al. Persistent Cryptococcus neoformans pulmonary infection in the rat is associated with intracellular parasitism, decreased inducible nitric oxide synthase expression, and altered antibody responsiveness to cryptococcal polysaccharide. Infection and Immunity. 2000;**68**(2):832-838

[147] Lee SC et al. Cryptococcus neoformans survive and replicate in human microglia. Laboratory Investigation; A Journal of Technical Methods and Pathology. 1995;**73**(6): 871-879

[148] Feldmesser M, Tucker S, Casadevall A. Intracellular parasitism of macrophages by Cryptococcus neoformans. Trends in Microbiology. 2001;**9**(6):273-278

[149] Woods JP et al. Pathogenesis of Histoplasma capsulatum. Seminars in Respiratory Infections. 2001;**16**(2): 91-101

[150] Porta A, Maresca B. Host response and Histoplasma capsulatum/ macrophage molecular interactions. Medical Mycology. 2000;**38**(6):399-406

[151] Strasser JE et al. Regulation of the macrophage vacuolar ATPase and phagosome-lysosome fusion by Histoplasma capsulatum. The Journal of Immunology. 1999;**162**(10):6148-6154

[152] Eissenberg LG, Goldman WE, Schlesinger PH. Histoplasma capsulatum modulates the acidification of phagolysosomes. Journal of Experimental Medicine. 1993;**177**(6): 1605-1611

[153] Eissenberg LG, Goldman WE. Histoplasma capsulatum fails to trigger release of superoxide from macrophages. Infection and Immunity. 1987;**55**(1):29-34

[154] Kozel TR. Complement and its role in fungal diseases. In: Human Fungal Pathogens. Springer; 2004. pp. 193-205

[155] Collette JR, Lorenz MC. Mechanisms of immune evasion in fungal pathogens. Current Opinion in Microbiology. 2011;**14**(6):668-675

[156] Zipfel PF, Skerka C. Complement regulators and inhibitory proteins. Nature Reviews Immunology. 2009;**9**(10):729-740

[157] Speth C et al. Complement and fungal pathogens: An update. Mycoses. 2008;**51**(6):477-496

[158] Speth C, Rambach G. Complement attack against Aspergillus and corresponding evasion mechanisms.

Interdisciplinary Perspectives on Infectious Diseases. 2012;**2012**

[159] Luo S et al. Complement and innate immune evasion strategies of the human pathogenic fungus Candida albicans. Molecular Immunology. 2013;**56**(3): 161-169

[160] Speth C et al. The role of complement in invasive fungal infections. Mycoses. 2004;**47**(3-4):93-103

[161] Kozel TR. Activation of the complement system by pathogenic fungi. Clinical Microbiology Reviews. 1996;**9**(1):34-46

[162] Hector R, Yee E, Collins M. Use of DBA/2N mice in models of systemic candidiasis and pulmonary and systemic aspergillosis. Infection and Immunity. 1990;**58**(5):1476-1478

[163] Tsai H-F et al. The developmentally regulated alb1 gene of Aspergillus fumigatus: Its role in modulation of conidial morphology and virulence. Journal of Bacteriology. 1998;**180**(12): 3031-3038

[164] Meri T et al. The yeast Candida albicans binds complement regulators factor H and FHL-1. Infection and Immunity. 2002;**70**(9):5185-5192

[165] Vogl G et al. Immune evasion by acquisition of complement inhibitors: The mould Aspergillus binds both factor H and C4b binding protein. Molecular Immunology. 2008;**45**(5):1485-1493

[166] Behnsen J et al. The opportunistic human pathogenic fungus Aspergillus fumigatus evades the host complement system. Infection and Immunity. 2008;**76**(2):820-827

[167] Mastellos D, Lambris JD. Complement: More than a 'guard'against invading pathogens? Trends in Immunology. 2002;**23**(10):485-491

[168] Ricklin D, Lambris JD. Complement in immune and inflammatory disorders: Pathophysiological mechanisms. The Journal of Immunology. 2013;**190**(8): 3831-3838

[169] Morgan BP, Harris CL. Complement, a target for therapy in inflammatory and degenerative diseases. Nature Reviews Drug Discovery. 2015;**14**(12):857-877

[170] Reis ES et al. Applying complement therapeutics to rare diseases. Clinical Immunology. 2015;**161**(2):225-240

Chapter 5

External Signal-Mediated Overall Role of Hormones/Pheromones in Fungi

Khirood Doley, Susan Thomas and Mahesh Borde

Abstract

The communication *via* signaling of chemicals is perhaps one of the earliest forms of communications. The most commonly known interspecific chemical substance such as pheromones is often known to engage in the attraction of mates in insects. Hence, the sensing of environmental and interindividual communication *via* pheromone systems is fundamental to most organisms that help in guiding the interactional behavior, development, and overall physiological activities. Likewise, the role of pheromones is revealed in fungal species in terms of their role in several cellular activities. The role of pheromones in fungi has been largely unexplored. However, there are few fungal hormones/pheromones such as sirenin, trisporic acid, antheridiol, oogoniol, and peptide hormone in yeast that were documented. Further studies are still underway for their significance in the biology of fungi as a whole and implications they might have on the overall ecosystem. In this chapter, we discuss various progresses made in understanding pheromone related to mating in kingdom fungi and the role of pheromone receptors.

Keywords: fungal hormones, pheromones, MAP kinase, signaling

1. Introduction

One of the largest and diverse kingdoms in eukaryotes is kingdom fungi, which may consist of 2.2–3.8 million species (approx.), and most importantly of which many are undergoing characterization, and it is well known to be present in several phyla that show important traits morphologically [1]. The fungal kingdom exhibits ubiquitous nature in our environment that plays a key role in the existence of life on earth as many species are directly or indirectly linked to in terms of several fields such as agricultural or industrial, and therefore, the implication is negatives as well positive for overall well-being of human and plant health [1]. The kingdom fungi are considered to be an ancient group of approximately 3.5 billion years old, which are comprised of large and diversely grouped organisms continuously updated with the discovery of new species yearly but the estimated number is around 1.5 and 7.1 million species [2–4], of which certain groups of fungi have been associated as pathogens for their ability to grow on humans, animals, or plants but there exists a helpful beneficial or mutual role also [5–9]. So far, the species that belong to this kingdom may include rusts, molds, lichens, smuts mushrooms, and yeasts.

By the large, the fungi have been relevant to their ability for undergoing the phenomena of secondary metabolism where secondary metabolites (SMs) as bioactive compounds are synthesized which demonstrates properties such as mutagenic, cytotoxic, immune-suppressants, antibiotics, and carcinogenic. In addition, various other SMs also have been shown to contribute to the interaction between host-plant and in resistances such as induced systemic and systemic acquired [10]. Both beneficial and harmful SMs have been reported from fungal species such as well-known antibiotic penicillin from *Penicillium*, sterigmatocystin, aflatoxin, and gliotoxin from *A. nidulans, A. flavus,* and *A. fumigates* [11, 12].

As far as sexual development in fungi is concerned, it is considered to be initiated by the fusion of haploid cells that are morphologically not distinguishable at all. de Bary (1981) was the first one who reported the occurrence of sex hormones in fungi such as *Achlya*. Later on, gradually various works on it revealed diffusible substance that plays a specific role during sexual reproduction in fungi.

Nonetheless, they can fuse due to differences in mating types. In most cases, under mating types, it undertakes overall particular activities of the cell that are essential for conversion from haploid to the diploid stage and meiosis. The genetic information determines the mating type and is supposedly present at the mating-type locus. So far, various systems have been evolved to make certain continuation of different sexes of either two or multiple mating types. In various fungi, diffusible peptide mating factors are mediated by particular cell recognition and fusion. These peptides act in a very similar way with the secreted chemical substances of insects and mammals especially in very low doses that elicit certain responses related to mating. Hence, it can be termed pheromones in a very similar way as in insects and mammals. The very first evidence of peptide pheromones came into existence from the observations when it was found that a certain diffusible substance was acting from a distance with an effect on cell-type specificity [13]. Hence, the occurrence and study of these pheromones have paved the way to study the varied fields of protein modification and their trafficking, signal transduction, ligand-receptor interactions, and cell cycle regulation. For this, the yeast *Saccharomyces cerevisiae* proved to be the best model organism due to its genetic, pheromone production response and molecular mechanisms involved in it. Furthermore, several studies have been undertaken in several species of fungi by use of several methods that may include the study of genetics, plant secondary metabolites, etc. for the occurrence of sexual reproduction and its related signaling cascades [10, 14].

Because the majority of fungi are non-motile, therefore, the property of responding to external cues such as signals helps in polarized growth especially in filamentous fungi for facilitating them in search for a continuous supply of essential nutrients and mating partners that prove to be crucial for their survival and abundance, while the mechanism in which external signals mediate in the process of polarization has marked commonalities with the polarization event during the processes such as mitotic division, budding, or fission. Thus, if any new substance is introduced into the outer membrane, then it may result in the growth in the polarized manner in the case of fungi using the secretory pathway and associated re-modeling of the cell wall. This type of growth had been reported to occur *via* internal as well as external signals such as cell cycle progression and environmental changes or probable occurrence of peptide mating pheromone, respectively. Most importantly, when opposite mating partners are exposed to the effects of pheromone, it has been found that several expressions in genes are observed along with the arrest of the cell cycle [15].

It has been extensively documented that polarized growth exists in several fungal species [16, 17]. Therefore, the biology-related mechanism present in pheromone-mating types of interactions in fungi kingdom may present helpful insights into the evolution of mating mechanisms, which will ultimately serve in understanding the not-so-studied models for sexual eukaryotes. In this chapter, we will not concentrate our views on chemotropism or cell fusion, which already have been extensively reviewed recently.

2. Sex hormone types

As far as endogenous hormone systems are concerned in fungi, it has very significant homologies for animals [18]. According to Machlis in 1972 [19], the sex hormones are may be classified into three following types *viz.*, erotactins, erotropins, and erogens where erotactin functions as a sexual hormone for attracting motile gametes, erotropin plays role in the induction of chemotropic growth of sexual structures, and lastly erogen role is to control the induction and differentiation of sexual structures. Nonetheless, the sex hormone may carry out more than one function for instance in *Achlya*, where it was found that it not only helps in controlling the overall development of sexual structures but also helps in determining the direction of the sexual organs.

3. Some common fungal sex hormones

Even though a large number of sex hormones have been investigated, but only a few of them have been characterized chemically or widely investigated, which are *viz.*, sirenin, trisporic acid, antheridiol, and oogoniol.

3.1 Sirenin

Among many fungal sex hormones, sirenin became the first known fungal sex hormone that was a sperm-attracting hormone and later on, it was classified according to its chemical composition. It has the basic property of female gamete that helps in attracting male gametes in genus *Allomyces*. It was first demonstrated in 1958 by Leonard Machlis [20], and organic chemists helped in its purification for its structural determination by 1968 as empirical formula $C_{15}H_{24}O_2$ with a molecular weight of 236. Consequently, further research has been carried out single-handedly on sirenin almost entirely by Jeffrey Pommerville. In his works, it was shown that male gametes helped in releasing a hormone that complements to sirenin, *parisin* which is attributed to attracting female gametes [21]. It resulted in the demonstration that there are parallels in the system of *Allomyces* hormone for several others that are present due to specific male as well as female hormones. But, unfortunately thereafter hardly any significant work has been carried out on *Allomyces*.

3.2 Trisporic acid

After the report of the first female sex hormone in form of sirenin in 1958, the last decade of the twentieth century saw a discovery of metabolite as trisporic acid. Trisporic acid is reported to be a sex hormone that has been isolated from *Blakeslea trispora* and *Mucor mucedo*, and it was shown to play an active role in sexual reproduction of various members of the order Mucorales. Also, it was found that trisporic acid caused significant carotene upregulation in the species of *B. trispora*. Afterward, in the sexual reproduction of *Mucor mucedo*, it was found that the hormone that was responsible for the process of gametangial conjugation that results in the production of zygospore was trisporic acid. Trisporic acid is an unsaturated and oxygenated form of trimethyl cyclo-hexane and there are three kinds of trisporic acid such as A, B, or C among them C is known to play chief role as a sex-hormone, afterward, trisporic acid B comes in terms of activity, followed by trisporic acid A with least activity. In the case of heterothallic mycelia, trisporic acid B and C have been found to stimulate the zygosphore developments and this particular hormone is produced only when the mycelia of (+) and (−) strains grow in a normal continuous diffusable medium. The trisporic acid hormone synthesized in (−) strain encourages the development of pro-gametangium in (−) strains or the other way round. The empirical formula of trisporic acid is C_{18} H_{26} O_4 with a molecular weight of 306. The revelation where sexual involvements are concerned which is under the control hormones in Mucorales group extends comprehensively over many years and workers from various countries are actively got involved.

3.3 Antheridiol and Oogoniol

During the early nineteenth century, John Raper discovered the hormone known as antheridiol which was initially termed as hormone A when he was studying the mode of mating in the oomycete water mold *Achlya*. It is demonstrated that the hormone antheridiol is responsible for several types of reactions such as antheridial hyphae initiation on the male plant, stimulation of antheridial hyphae chemotropic way, male hyphae stimulation for the oogoniol production, and their role in antheridia delimitation. The hormone was retrieved from the female was obtained by Raper and the chemist Haagen-Smit in a highly concentrated state was earlier shown to induce the development of antheridia in the male by the early twentieth by Raper. The hormone oogoniol is synthesized by male hyphae of *Achlya ambisexualis* but only when antheridiol is present. However, it was reported that oogoniol may be synthesized as well by some hermaphrodite strains exclusive of antheridiol stimulus [22]. McMorris [23] and his coworkers found that two crystalline compounds that possessed hormone B activity have been isolated from culture filtrates of *Achlya heterosexualis* which received the name of oogoniol-1 and oogoniol-2. Hence, the hormone that stimulates the development of oogonium on female hyphae came into existence as oogoniol that is a crystalline steroid with 500 as molecular weight.

3.4 Yeast a-factor and alpha-factor

In *S. cerevisiae*, mating-type factors are peptide hormones called a and alpha pheromones and they have specific receptors on the cell surface. These pheromones binding to specific receptors on opposite mating-type cause G1 arrest in the cell cycle which is the same stage that is required for nuclear fusion. Investigating the effects on the cell cycle by the responses generated by signal transduction to the nucleus *via* extracellular pheromones seems to be a better prospect. In *S. cerevisiae*, the receptor family that it belongs to is the large family of receptors known as G-protein-coupled serpentine seven-trans-membrane (GPCR) receptor and it has been involved in studying these receptors as a useful model system for the investigation of complex signal transduction cascade. And the enzymes that are required for this signal transduction have very significant implication for the eukaryotic protein such as RAS oncoproteins because it is proofing useful in various forms of cancer. Hence, studying yeast mating has therapeutic value as it may provide novel tumor-suppressing agents. Nevertheless, the marked evidences of fungal pheromones seem to be widespread and it not only have a significant role in cell-to-cell recognition but also in post-fusion events *viz.*, induction of meiosis and maintenance of the filamentous state in some species of fungi [24].

4. Basidiomycetes pheromone signaling

If we have a look at the system of mating in basidiomycetes, it consists of haploid monokaryotic that induces dikaryotic stages. The monokaryotic mycelium consists of nuclei with one genetic type; therefore, the terminology homokaryon is derived. From the haploid spores, the mycelium grows, which may contain one nucleus. And when genetically different types of homokaryons mate, then dikaryon is formed. Generally, the tetrapolar system regulates the mating in the mushroom-forming species, which consists of genetic complexes namely *A* and *B*, which may not be linked with each other. The condition of tetrapolar reveals that there may be possibilities that are four in number as far as mating interactions are concerned involving haploid strains. And full compatible interaction may take place between mates when both genetic complexes have different specificities as compared to different allelic specificities and two semi-compatible interactions have been observed to take place when development is regulated by *A*- or *B* due to differences present in either complex. As far as basidiomycetes are concerned, both kinds of mating behavior are observed. The induction of sexual development during mating has to be responsive to binding of ligand on pheromone receptor followed by an ensuing signal transduction pathway, which will ultimately leads to dikaryotization. Hence, one of the chief functions in a system where pheromone and receptors are involved is to bring about the gene expressions of encoded proteins that are involved in attracting and subsequently directing mates to grow toward each other.

Hence, the interaction between pheromone and receptor is believed to be significantly appropriate for proceedings of fusion and most particularly when there is the presence of shared exchanges and migration of nuclei among the mating partners [25]. In addition, the evidences suggest that there is hardly any significant correlation between the strength of responses in species or its genetic distance from pheromone source sequence but due to influencing conditions or differences in development a species may be either weak or strong responder [26].

So, during a response to pheromone and nuclear migration in the fungal species of *S. cerevisiae*, it has been observed that a mating-specific Gα, Gpa1, has been found to interact with kinesin-14 (Kar-3) that is a minus-end-directed microtubule-associated motor [27]. In addition, due to this interaction nuclear migration induced by pheromone is regulated toward shmoo tip. After pheromone treatment, Kar3 immunoprecipitates the Gpa1, thereby demonstrating interactions among protein–protein complexes. Finally, at the shmoo tip visualization of Gpa1 and Kar3 occurs. It was regarded that the positive association of Kar3 with Gpa1 gets affected when utilization of mutant Gpa1 is undertaken to the shmoo tip. The dynamics and orientation of microtubules also have a significant association with Gpa1. It was concluded that Gpa1 was considered to provide an externally regulated position determinant for the anchorage of Kar3 [27]. Even though, the conserved pheromone/receptor system and the presence of different regulations, the *de novo in silico* discovery of proteins similar to a receptor, and interactions of various intracellular signal transduction pathways build the perceptive of the origin as well as the functionality of the pheromone receptor system in highly complex systems of agaricomycetes a challenge to undertake [28].

As far as detection of external stimuli is concerned in eukaryotic organisms, there are arrangements in terms of signaling transduction pathways that are employed for detections [29]. The signal transduction pathway occurs *via* mitogen-activated protein kinases (MAPKs) that comprises of Ser/Thr protein kinases which helps in converting the extracellular stimuli into several downstream cellular responses. And it has been reported that since ancient times the MAPKs signal transduction pathway is extensively utilized in many physiological processes throughout evolution.

Hence, we can mention that MAPK pathway involves well-conserved signal transduction cascade in eukaryotes [30]. The signaling pathway plays a significant role in response to several factors such as growth factors, mitosis, gene expression, cytokine regulation, motility, metabolism, cell death, cellular stressors, differentiation, and pheromones [29]. In mammals, MAPKs are 14 in number and have been characterized into 7 groups. Generally, MAPK pathways consist of kinases that are MAP3K (MAPKKK), MAP2K (MAPKK), and MAPK that upon stimulus detection become co-localized and subsequently allow sequential phosphorylation and activation. The MAPKKKs are Ser/Thr protein kinases that get activated due to phosphorylation results into their interaction with small proteins that are GTP-binded that belongs to Ras/Rho family. Hence, activation of MAPKKK happens and it directs phosphorylation, which brings about the MAPK activity *via* phosphorylation of Thr and Tyr residues within a conserved motif known as Thr-X-Tyr located in the kinase domain [29]. Thus, this pathway consists of several described adaptors, docking, and scaffold proteins that are occupied in the overall regulation of MAPK cascade of signaling. In these, the scaffolds are considered to be the largest, a multi-domain protein. The scaffold protein is provided as a physical platform that has the capability of binding several members of a MAPK pathway that regulates in allocating in the regulation of localization of kinase, complex assembly, and transcriptional factors for signal propagation toward nucleus for respective expression [31].

Nevertheless, in fungi, the growth and development and several other processes require a wide range of signal transduction pathways [31–33]. Therefore, various MAPK-involved pathways have been reported in the biological regulations concerned in fungal species. So far, all MAPK pathways in *S. cerevisiae* have been defined genetically. But the pathway related to mating was the first MAPK module to be defined. *S. cerevisiae* has been reported to have five MAPK pathways (Fus3, Kss1, Hog1, Mpk1, and Mpl1) wherein each regulates separate cellular processes [34].

Despite very few information available on the signaling mechanisms that are involved in fungal development, still among various known pathways, especially in eukaryotes sexual development occurs *via* the MAPK pathway, which is widely studied during pheromone signaling [35]. In addition, since it came into existence, it is known to be highly conserved in the fungi in terms of orthologous pheromone module MAPK pathways. In the review of Frawley and Bayram [31], where they specifically mentioned the role of a module that involves pheromone as a foremost signaling in filamentous fungal species *viz.*, *A. nidulans*, *A. flavus*, and *A. fumigates*.

For instance, it has been shown in the case of filamentous growth in *S. cerevisiae* *via* MAPK pathway in response to pheromones where Ste7, Ste11, Ste20, and Kss1 kinases and the Ste12 transcription factor acted as several components [36, 37]. However, in diploid cells, for the filamentous growth pheromones, pheromone receptors and subunits of the pheromone-activated heterotrimeric G protein are not necessarily expressed in case of diploid cells (161). The signaling, in particular, involves different specializations *viz.*,

i. Activation of the MAP kinase cascade by the beta-gamma subunits of the pheromone-activated heterotrimeric G protein during mating of haploid cells. There exist various mechanisms that involve Cdc42, Ras2, and 14–3-3 proteins Bmh1 and Bmh2 during growth in filamentous way and this ultimately leads to the activation of the MAP kinase pathway [38, 39].

ii. During mating the scaffold protein Ste5 helps in securing the components of the MAP kinase cascade; however, during filamentous growth Ste5 is not vital and the role of scaffolding may be played by another protein. In addition, Spa2 protein is suggested to interact with the MAP kinase cascade components as the scaffold during filamentous growth [40].

iii. The Fus3 and Kss1 kinases differ in the MAP kinase of *S. cerevisiae* where Fus3 specifically regulates mating and invasive growth is inhibited while Kss1 is comprehensively assisted in the regulation of invasive and filamentous growth [41]. Furthermore, during filamentous growth Kss1 role has been positive and negative regulation and it helps in preventing repression of Ste12 by the proteins Dig1 and Dig2 [42–44].

iv. In the last specialization, Ste12 interacts with the Mcm1 protein during mating and transcription is activated, which eventually leads to gene expression of pheromone-responding features, while heterodimer formation occurs in diploid cells by Ste12 accompanied by Tec1 that leads to activation of transcription related genes along with filamentation-responding features [37]. Thus, in such a way, Ste12 protein may give way for two different patterns of suitable transcriptional responses in haploid and diploid cells.

In the case of *A. flavus* pheromone module, there is the presence of kinases and SteD and the process of dimerization occurs at hyphal tip as MkkB-MpkB where it interacts in the cytoplasm as SteC-SteD dimer and a tetrameric complex is formed. From this tetrameric complex, phosphorylation of MpkB occurs and it enters the nucleus and respective regulation of transcriptional factors occurs. Furthermore, at the hyphal tip HamE localization is also observed for respective regulations. However, the exact mechanisms that are involved in its regulation are still subjected to further research.

5. Conclusions

The signal transduction cascade in the fungal kingdom which is highly conserved occurs not only by the element of pheromone but also by both secondary metabolism and pathogenicity present in various fungal pathogens. But still very few investigations have been carried out about the pathways that involve the protein complexes for signaling to happen in the fungal kingdom.

The fungal pheromones are generally secreted by the opposite mating cells to stimulate the production of the opposite sex organs. The fungal pheromone such as sirenin is produced by female gametes of Allomyces is used for attraction of male gametes and fusion. Trisporic acid reported from Zygomycetes fungi is responsible for zygotrophism and development of progametangium. Antheridiol and oogoniol hormones are produced by *Achlyabisexualis*, and vegetative hyphae of female strain-produced antheridiol are responsible for the development of antheridial hyphae on male thallus. Then, the male hyphae-produced oogoniol causes the initiation of oogonial hyphae on female thallus. In yeast, peptide hormone a and alpha-factor have specific receptors present on the opposite mating types. These peptide hormones bind to receptors that are specific to the cell surface of opposite mating type through GTP-binding protein causes the production of agglutinin of recipient cell and stops the G1 stage of the cell cycle. In basidiomycetes fungi different mating homokaryons are having either A or B genetic alleles and also have two semi-compatible interactions been observed in it leading to dikaryotization.

Even though there are evidences of the signaling pathway taking part in the regulation of various fungal progression that may be vegetative in nature, sporulation asexually, and sexual reproduction but the evidences concerning the requisite stimulation to activate these pathways or transcriptional regulation in the nucleus of filamentous fungi especially are meager. Therefore, there are still more prospects in the field of complex signal transduction pathways present in fungi and the mechanisms *via* in which certain genes are regulated utilizing the pheromone module. Hence, if in near future our researchers are able to characterize the vital regulators in the development of the fungal kingdom, then it could help in the sustenance of the global population by reducing food production loss by preventing various fungal crop diseases.

Conflict of interest

On behalf of all authors, the authors declare no competing and conflict of interest.

Author details

Khirood Doley[1]*, Susan Thomas[2] and Mahesh Borde[1]

1 Department of Botany, Savitribai Phule Pune University, Pune, India

2 Department of Botany, Spicer Adventist University, Pune, India

*Address all correspondence to: khirood_doleys@yahoo.com

References

[1] Hawksworth DL, Lücking R. Fungal diversity revisited: 2.2 to 3.8 million species. Microbiology Spectrum. 2017;5(4). DOI: 10.1128/microbiolspec. FUNK-0052-2016

[2] Blackwell M. The Fungi: 1, 2, 3...5.1 million species? American Journal of Botany. 2011;**98**:426-438

[3] Parfrey LW, Lahr DJG, Knoll AH, Katz LA. Estimating the timing of early eukaryotic diversification with multigene molecular clocks. Proceedings of the National Academy of Sciences of the United States of America. 2011;**108**:13624-13629

[4] Dornburg A, Townsend JP, Wang Z. Maximizing power in phylo- genetics and phylogenomics: A perspective illuminated by fungal big data. Advances in Genetics. 2017;**100**:1-47

[5] Voyles J, Young S, Berger L, Campbell C, Voyles WF, Dinudom A, et al. Pathogenesis of chytridiomycosis, a cause of catastrophic amphibian declines. Science. 2009;**326**:582-585

[6] Evans HC, Elliot SL, Hughes DP. Hidden diversity behind the zombie-ant fungus *Ophiocordyceps unilateralis* four new species described from carpenter ants in *Minas Gerais*, Brazil. PLoS One. 2011;**6**:e17024. DOI: 10:1371/journal. pone.0017024

[7] Hughes DP, Anderson S, Hywel-Jones NL, Himaman W, Billen J, Boomsma JJ. Behavioral mechanisms and morphological symptoms of zombie ants dying from fungal infection. BMC Ecology. 2011;**11**:13. DOI: 10.1186/ 1472-6785-11-13

[8] Ziaee A, Zia M, Goli M. Identification of saprophytic and allergenic fungi in indoor and outdoor environments. Environmental Monitoring and Assessment. 2018;**190**:574. DOI: 10.1007/s10661-018-6952-4

[9] Begum N, Qin C, Ahanger MA, Raza S, Khan MI, Ashraf M, et al. Role of Arbuscular Mycorrhizal fungi in plant growth regulation: Implications in abiotic stress tolerance. Frontiers in Plant Science. 2019;**10**:1068. DOI: 10.3389/ fpls.2019.01068

[10] Pang Z, Chen J, Wang T, Gao C, Li Z, Guo L, et al. Linking plant secondary metabolites and plant microbiomes: A review. Frontiers in Plant Science. 2021;**12**:621276. DOI: 10.3389/fpls.2021. 621276

[11] Bills GF, Gloer JB. Biologically active secondary metabolites from the fungi. Microbiology Spectrum. 2016;**4**(6). DOI: 10.1128/microbiolspec.FUNK-0009-2016 PMID: 27809954

[12] Amaike S, Keller NP. *Aspergillus flavus*. Annual Review Phytopathology. 2011;**49**:107-133. DOI: 10.1146/annurev-phyto-072910-095221 PMID: 21513456

[13] Levi JD. Mating reaction in yeast. Nature. 1956;**177**:753-754

[14] Ni M, Feretzaki M, Sun S, Wang X, Heitman J. Sex in fungi. Annual Review of Genetics. 2011;**45**:405-430. DOI: 10.1146/annurev-genet-110410-132536

[15] Wallen RM, Perlin MH. An overview of the function and maintenance of sexual reproduction in Dikaryotic fungi. Frontiers in Microbiology. 2018;**9**:503. DOI: 10.3389/ fmicb.2018.00503

[16] Fischer R, Zekert N, Takeshita N. Polarized growth in fungi – interplay between the cytoskeleton, positional markers and membrane domains. Molecular Microbiology. 2008; **68**:813-826

[17] Takeshita N, Manck R, Grün N, de Vega SH, Fischer R. Interdependence of the actin and the microtubule cytoskeleton during fungal growth. Current Opinion in Microbiology. 2014;**20**:34-41. DOI: 10.1016/j.mib.2014.04.005 Epub 2014 May 27. PMID: 24879477

[18] Gooday GW, Adams DJ. Sex hormones and fungi. Advances in Microbial Physiology. 1992;**34**:69-145

[19] Machlis L. The coming of age of sex hormones in plants. Mycologia. 1972;**64**:235-247 PMID: 4553262

[20] Machlis L. Evidence for a sexual hormone in *Allomyces*. Physiologia Plantarum. 1958;**11**:181-192

[21] Pommerville JC, Strickland JB, Romo D, Harding KE. Effects of analogues of the fungal sex pheromone sirenin on male gamete motility in *Allomyces macrogynus*. Plant Physiology. 1988;**88**:139-142

[22] Barksdale AW, Lasure LL. Production of hormone B by Achlya heterosexualis. Applied Microbiology. 1974;**28**:544-546

[23] McMorris TC, Seshadri R, Weihe GR, Arsenault GP, Barksdale AW. Structures of oogoniols1, −2, and −3, steroidal sex hormones of the water mold *Achlya*. Journal of the American Chemical Society. 1975;**97**:2544-2555

[24] Bölker M, Kahmann R. Sexual pheromones and mating responses in fungi. The Plant Cell. 1993;**5**:1461-1469. DOI: 10.1105/tpc.5.10.1461

[25] Raudaskoski M. The relationship between *B* mating type genes and nuclear migration in *Schizophyllum commune*. Fungal Genetics and Biology. 1998;**24**:207-227

[26] Xu L, Petit E, Hood ME. Variation in mate-recognition pheromones of the fungal genus *Microbotryum*. Heredity. 2016;**116**:44-51. DOI: 10.1038/hdy.2015.68

[27] Zaichick SV, Metodiev MV, Nelson SA, Durbrovskyi O, Draper E, Cooper JA, et al. The mating-specific Gα interacts with a kinesin-14 and regulates pheromone-induced nuclear migration in budding yeast. Molecular Biology of Cell. 2009;**20**:2820-2830

[28] Raudaskoski M, Kothe E. Basidiomycete mating type genes and pheromone signaling. Eukaryotic Cell. 2010;**9**:847-859. DOI: 10.1128/EC.00319-09 Epub 2010 Feb 26. PMID: 20190072; PMCID: PMC2901643

[29] Cargnello M, Roux PP. Activation and function of the MAPKs and their substrates, the MAPK-activated protein kinases. Microbiology and Molecular Biology Reviews. 2011;**75**:50-83. DOI: 10.1128/MMBR.00031-10

[30] Widmann C, Gibson S, Jarpe MB, Johnson GL. Mitogen - activated protein kinase: Conservation of a three-kinase module from yeast to human. Physiological Reviews. 1999;**79**:143-180

[31] Frawley D, Bayram O. The pheromone response module, a mitogen-activated protein kinase pathway implicated in the regulation of fungal development, secondary metabolism and pathogenicity. Fungal Genetics and Biology. 2020;**144**:103469. DOI: https://doi.org/10.1016/j.fgb.2020.103469

[32] Lengeler KB, Davidson RC, D'souza C, Harashima T, Shen WC, Wang P, et al. Signal transduction cascades regulating fungal development and virulence. Microbiology and Molecular Biology Reviews. 2000;**64**:746-785. DOI: 10.1128/MMBR.64.4.746-785.2000

[33] Elramli N, Karahoda B, Sarikaya-Bayram Ö, Frawley D, Ulas M, Oakley CE, et al. Assembly of a

heptameric STRIPAK complex is required for coordination of light-dependent multicellular fungal development with secondary metabolism in *Aspergillus nidulans*. PLoS Genetics. 2019;**18**(15):e1008053. DOI: 10.1371/journal.pgen.1008053 PMID: 30883543; PMCID: PMC6438568

[34] Qi M, Elion EA. MAP kinase pathways. Journal of Cell Science. 2005; **118**:3569-3572

[35] Bardwell L. A walk-through of the yeast mating pheromone response pathway. Peptides. 2005;**26**:339-350

[36] Lo WS, Dranginis AM. The cell surface flocculin Flo11 is required for pseudohyphae formation and invasion by *Saccharomyces cerevisiae*. Molecular Biology of the Cell. 1998;**9**:161-171

[37] Madhani HD, Fink GR. Combinatorial control required for the specificity of yeast MAPK signaling. Science. 1997;**275**:1314-1317

[38] Mosch HU, Fink GR. Dissection of filamentous growth by transposon mutagenesis in *Saccharomyces cerevisiae*. Genetics. 1997;**145**:671-684

[39] Roberts R, Mosch HU, Fink GR. 14-3-3 proteins are essential for RAS/MAPK cascade signaling during pseudohyphal development in *S. cerevisiae*. Cell. 1997;**89**:1055-1065

[40] Roemer T, Vallier L, Sheu YJ, Snyder M. The Spa2-related protein, Sph1p, is important for polarized growth in yeast. Journal of Cell Science. 1998;**111**:479-494 PMID: 9443897

[41] Cook JG, Bardwell L, Thorner J. Inhibitory and activating functions for MAPK Kss1 in the *S. cerevisiae* filamentous-growth signalling pathway. Nature. 1997;**390**:85-88

[42] Bardwell L, Cook JG, Voora D, Baggott DM, Martinez AR, Thorner J. Repression of yeast Ste12 transcription factor by direct binding of unphosphorylated Kss1 MAPK and its regulation by the Ste7 MEK. Genes and Development. 1998;**12**:2887-2898

[43] Madhani HD, Fink GR. The riddle of MAP kinase signalling specificity. Trends in Genetics. 1998;**14**:151-155

[44] Cook JG, Bardwell L, Kron SJ, Thorner J. Two novel targets of the MAP kinase Kss1 are negative regulators of invasive growth in the yeast *Saccharomyces cerevisiae*. Genes and Development. 1996;**10**:2831-2848

Chapter 6

Glomalin Arbuscular Mycorrhizal Fungal Reproduction, Lifestyle and Dynamic Role in Global Sustainable Agriculture for Future Generation

Kamal Prasad, Agam Khare and Prateek Rawat

Abstract

Glomalin, a type of glycoprotein produced by arbuscular mycorrhizal fungi in the phylum *Glomeromycota*, contributes to the mitigation of soil degradation. Moreover, AM fungi and glomalin are highly correlated with other soil physico-chemical parameters and are sensitive to changes in the environment; also, they have been recommended for monitoring the recovery of degraded soil or stages of soil degradation. AM fungi are commonly known as bio-fertilisers. Moreover, it is widely believed that the inoculation of AM fungi provides tolerance to host plants against various stressful situations like heat, salinity, drought, metals and extreme temperatures. AM fungi, being natural root symbionts, provide essential plant inorganic nutrients to host plants, thereby improving growth and yield under unstressed and stressed regimes. The role of AM fungi as a bio-fertiliser can potentially strengthen plants' adaptability to changing environment. They also improve plant resilience to plant diseases and root system development, allowing for better nutrient absorption from the soil. As a result, they can be utilised as both a biofertilizer and a biocontrol agent. Present manuscript represents the potential of AM fungi as biostimulants can probably strengthen plants' ability to change the agriculture system for green technology.

Keywords: Glomalin, AM fungi, reproduction, Symbiosis, biocontrol agent

1. Introduction

Glomalin levels are high in soils and are linked to aggregate water stability. Glomalin contains carbon and hence contributes a significant amount of carbon to the terrestrial carbon pool. Stabilisation of aggregates, on the other hand, likely increases the effect of glomalin in soils by protecting carbonaceous molecules from degradation within aggregates. Because of the symbiotic relationship that occurs between plants and glomalin producers, AM fungus, higher atmospheric CO_2 can lead to increased glomalin production. The agroecosystem's management strategies have an impact on glomalin concentrations in soils. Carbon storage is an important function of glycoprotein in soil. Glomalin is a rare molecule (protein) that has been

IntechOpen

difficult to study biochemically due to its resistance and complexity. Fungi could be a microscopic microorganism of the cluster eukaryotes that consists of yeasts, moulds, and mushrooms. These organisms are terribly little requiring a magnifier for thorough observation. They are globally plentiful and located in a very vast sort of habitat.

There are some beneficial fungus species in the hemisphere that have shaped civilization and fungi have had a significant impact on human and plant longevity. Plants began putting down roots in terrestrial habitats over 460 million years ago, and they were determined by a symbiotic fungus called mycorrhizae. AM fungi (Endomycorrhiza) are grouped into a monophyletic phylum, Glomeromycota, which includes all notable AM fungi and has coevolved with the majority of plants since then. Given mycorrhizae's long evolutionary history, it's not surprising that the mycorrhizal connection is found in more than 95% of all vascular plants, as AM fungi appear to lack host specificity. Plants and glycoprotein-producing fungi form a root endosymbiosis known as AM fungi (GPPF). It is the most widely distributed terrestrial plant symbiont, helping the plant absorb more water and mineral nutrients. Mycorrhizae share some primitive fungal traits, including the ability to form spores, a lack of diversity, the lack of sexual reproduction and the inability to thrive without a living host. The hemisphere is home to a wide variety of mycorrhizae. In forest plants, ectomycorrhizas rely on fungi surrounding the roots in a sheath (mantle) and a Hartig net of hyphae that extends into the roots between cells. The fungal companion could be from the Ascomycota, Basidiomycota, or Zygomycota families. Glomeromycota fungus creates vesicular-arbuscular contacts with AM fungi in a second type. AM fungi produce arbuscular cells, which penetrate root cells and serve as a conduit for metabolic exchanges between the fungus and the host plant. The arbuscules (small trees with a bushy appearance) have a bushy appearance. Orchids are dependent on a third type of mycorrhiza. Orchids are epiphytes with little seeds that require a lot of storage to survive germination and growth. Without a mycorrhizal companion, their seeds do not germinate (Basidiomycete). Once the seed's nutrients are spent, fungal symbionts help the orchid grow by delivering vital carbohydrates and minerals. Throughout their lives, a few orchids remain mycorrhizal connections. AM fungus is obligate biotrophs, meaning they only eat the products of their live hosts' photosynthesis. Fungi aren't usually specialised for their possible hosts, yet some plant species are more conducive to the growth of those fungi than others [1–4]. Fungi are among the most commonly found soil microorganisms on the globe, and they are related to plants such as angiosperms, gymnosperms and pteridophytes with roots, as well as the gametophytes of a few mosses, lycopods and Psilotalus, which do not have true roots [2]. According to numerous studies, AM fungi increase root absorptive area and, as a result, plant nutrition [5, 6], influence plant community succession [7], their fight [8, 9] and phenology [8] equalise the extent of nutrition of co-existing plants by forming hyphal bridges that transfer nutrients among them [10], and increase soil structure by binding sand grains into aggregates by ERH [11, 12]. Plants' tolerance to heavy metals [13–15], water stressors [16], pathogenic fungus, and nematodes was increased by AM fungi [17, 18]. The need for up to 20% of host photosynthate by AM fungus for establishment and maintenance is well understood [19, 20]. This manuscript focuses on the lifecycle and potential role of AM fungi as biofertilizers inside the regulation of plant growth, development, with improved nutrient uptake to a lower place disagreeable environment, overall crop improvement and changing universal sustainable agriculture for future generations and greening agriculture.

2. AM fungi upbringing

Intraradical hyphae (IRH) within the roots and extraradical hyphae (ERH) structures outside the roots are found in AM fungi. Arbuscules, vesicles and intraradical hyphae are among the IRH structures. The extraradical hyphal structures are spores, and the auxiliary cells are Gigaspora, Pacispora and Scutellospora members. The principal locations of nutrition exchange between a host plant and a fungal flora are haustoriums and arbuscules [4, 12, 13, 15, 20]. They are made up of cells in the internal root cortex (IRC) [1, 4, 21, 22] and are signs of active, lively and alive mycorrhizae. Arbuscules come in a variety of shapes and sizes, and their form is based on the common association of arbuscular fungi [23]. Arbuscules with cylindrical or slightly flared, slender trunks are produced by fungi of the genera Acaulospora, Archaeospora, Ambispora, Diversispora, Entrophospora, Glomus, Intrapora, Kuklospora, Pacispora and Paraglomus. Members of the genera Gigaspora and Scutellospora have large trunks and branches that taper abruptly at the tips. Globose, spherical, or ovoid, thin-walled vesicles are lipid and glycolipid storage organs [24]. Intercalary swelling in the root ends of intraradical hyphae produces AM fungal vesicles.

Glomus vesicles are mostly elliptical, but Acaulospora, Entrophospora and Kuklospora vesicles have a wide range of shapes and rarely feature knobs or concavities on their surface [23]. Members of the genera Gigaspora and Scutellospora never produce vesicles. Vesicles are rarely produced by members of the genera Archaeospora, Intraspora and Paraglomus. To boot, intercellular hyphae (ICH) in roots store materials and help transfer elements absorbed by extraradical hyphae from the soil to arbuscules or directly to the host plant's root cells [1, 4, 5]. Intraradical hyphae can be straight or have branches that form an H or Y shape. They may additionally form coils, whose frequency of incidence depends upon their position in a root and therefore the generic affiliation of the arbuscular fungous species [23].

In general, coils proliferate at access locations. Glomus species' intraradical hyphae are sometimes coiled within the other areas of an AM fungal root. Coils produced by other AM fungus genera, on the other hand, are occasionally abundant and evenly scattered, along with mycorrhizal roots. The degree of evenness of dispersion of roots among AM fungus, and hence the intensity of staining, varies as well. Members of the genera Ambispora, Archaeospora, Acaulospora, Diversispora, Entrophospora, Intraspora, Kuklospora and Paraglomus have patchy distributions of AM fungous structures, whereas mycorrhizae of the genera Gigaspora, Glomus, Pacispora and Scutellospora have a consistent distribution. The staining power of Ambispora, Archaeospora, Diversispora, Intraspora and Paraglomus mushrooms may be very faint to faint, Acaulospora, Entrophospora and Kuklospora fungi faint to moderate, Glomus fungi dark, Gigaspora, Pacispora and Scutellospora fungi extremely dark [25, 26]. The sub-phylum Glomeromycota of the phylum Mucoromycotina contains the bulk of AM fungus species [27]. Glomerales, Archaeosporales, Paraglomerales and Diversisporales are the four orders of AM fungi that make up this subphylum, which also includes twenty-five genera [28]. They are obligate biotrophs and ingest plant photosynthetic products [29] and lipids to perform their lifecycle [30]. AM fungi-mediated growth promotion is not solely by enhancing water and mineral nutrients uptake from the conterminous soil but to boot by means of safeguarding the plants from fungal pathogens [31, 32]. Therefore, AM fungi are essential endosymbionts taking part in an efficient role in plant productivity and therefore the functioning of the ecosystem for sustainable crop enhancement.

3. AM fungi paleobiology

AM fungi are thought to be an ancient symbiosis that began over a million years ago, based on paleobiological and molecular evidence. The symbiosis of AM fungus with terrestrial plants is widespread, implying that mycorrhizas were present in the ancestors of all contemporary universal living plants. This favourable relationship with plants may have aided the development of terrestrial plants. Wherever AM fungi are found, fossils of the first land plants have been found in the Rhynie chert from the lower Devonian period [32]. Colonised fossil roots had been ascertained in Aglaophyton foremost and Rhynia, which might be ancient plants possessing characteristics of vascular plants and bryophytes with primitive protostelic rhizomes [33]. The fossil arbuscules seem much similar to those of existing AM fungi [33]. Mycorrhizas from the Miocene show a vesicular morphology closely resembling that of present Glomerales. This preserved morphology can even to boot replicate the prepared accessibility of nutrients provided by the plant hosts in each fashionable and Miocene symbiosis [32]. However, it might be argued that the effectiveness of sign approaches is probable to have evolved since the Miocene, and this cannot be detected within the fossil record.

3.1 AM fungi molecular signal

The upward interest in AM fungal symbiosis and the improvement of sophisticated molecular techniques have resulted in a rapid improvement in the genetic signals. Wang et al. [34] studied plant genes including DMI1, DMI3, IPD3, which are involved in communication with associated fungi of the order Glomales. The phylogeny of these three genes has been proven to be congruent with the present phylogeny of land plants, and they can be sequenced from all major clades of modern land plants, including liverwort, the maximum basal group. This suggests that mycorrhizal genes must have existed in the common ancestor of land plants and must have been passed down vertically to colonised land plants [34].

4. AM fungi reproduction and lifecycle

AM fungi reproduce by forming spores at the ends of the hyphae. These thick-walled spores stayed underground for a long period of time. Spores of AM fungi can germinate and form hyphae with living hosts.

4.1 AM fungi pre-symbiosis

The amplification of AM fungi before root colonisation (RC), called pre-symbiosis, involves three stages, including spore germination, hyphae growth, host recognition, and appressorium formation.

4.1.1 AM fungal spore germination

The reproduction of AM fungal spores is usually carried out with the help of asexual spores. AM fungal spores are thick-walled, multinucleated, resting structures, especially at the end of continuous sporulating hyphae with mycorrhizal extraradical hyphae. Spore germination has nothing to do with plants because spores germinate *in vitro* (modified living roots) and *in vivo* experimental conditions with and without plants; however, with the help of host root exudates, the germination rate may be hyperbolic. AM fungal spores germinate under suitable soil substrate conditions, temperature, CO_2 concentration, pH value and phosphate condition (PC).

4.1.2 AM fungi hyphal development

Host root exudates were known as strigolactones, and hence the soil PC, control the development of AM fungal hyphae through the soil. Low PC levels in the soil promote hyphal growth (HG) and branching, as well as plant exudation of chemicals that regulate hyphal branching intensity [35, 36]. AM fungal hyphae produced in 1 mM P media had significantly less branching; however, the length of the germ tube and overall hyphal development are unaffected. Every hyphal development and branching of AM fungus has a stage of 10 mM P pent-up. This PC occurs in natural soil environments and may hence contribute to lower AM fungus invasion [35].

4.1.3 AM fungi host recognition

It has been shown that root exudates (RE) of AM fungal host plants grown in phosphorus-containing and non-phosphorus-containing liquid media can affect mycelial growth. *Gigaspora margarita* and *Glomus intraradices* spores grow on host exudate. Hyphae of AM fungi, compared with plant exudates injected with P, root exudates lacking P grow in large numbers and form tertiary branches. Among the highest concentrations of arbuscular branches, the AM fungal structure is formed by phosphorus exchange [35]. This allows the hyphal growth (HG) to grow closer to the roots of potential host plants; the spores of *Glomus mosseae* are separated from the roots of the host plant through an osmotic membrane, rather than separated from plants and dead plants, effectively becoming hyphae. When treated with host plants, the fungi penetrated the membrane and appeared continuously within 800 μm of the roots, but now they are no longer included in the preparation of non-host plants and dead plants [37].

4.1.4 AM fungi appressorium/infection structure

The hyphae of AM fungi encounter the root foundation of the host plant, forming appressorium or infectious structures in the root epidermis. From this structure, the hyphae can enter the parenchymal cortex of the host. AM fungi would really like no chemical signals from the plant to make the appressoria. AM fungi can form adherent cells on the cell wall of ghost cells, where the protoplasts are removed to prevent signal transmission between the fungus and the host plant. Hyphae do not invade cells in a similar manner and develop near the root cortex, which suggests that once attachments are formed, signal transmission between symbionts is necessary for similar increased growth as soon as appressoria [36].

4.2 AM fungal cell structure, metabolism and natural life

AM fungus is obligatory organism that must complete their life cycle and produce next-generation spores on living photosynthetic autotrophic hosts. AM fungi are spores that grow on the top of the hyphae and are fully asexual. The spores of the AM fungus grow on the outside or inside of the host root. AM fungal spores can germinate in vitro without a host plant when they come into contact with modified live roots. Spores develop and form a germination tube that extends through the soil until it finds a host root in the absence of live roots. AM fungal spores penetrate roots and develop between root cells or penetrate cell walls and grow inside root cells. Once the spore penetrates the root cell, arbuscular branches are formed. Arbuscules branches are tree-like subcellular structures used to exchange nutrients between AM fungi and related symbiotic plants. The hyphae in the soil may also exceed 100 meters per cubic centimetre [38]. This network of hyphae is designed to increase the absorption of important macro and micronutrients by plants, including N, P, K, Zn, Fe, S, Mn, Mg, Cu, and water.

4.3 AM fungi symbiosis

AM fungi form a highly branched structure in the parenchyma, which is used to exchange nutrients in the plant referred arbuscules [1, 4, 6, 39]. These are specific structures unique to AM fungi. Arbuscules are exchange points for replacing phosphorus, carbon, water and other nutrients [1, 4, 15, 18, 40, 41]. There are two types: Paris forms, which have hyphae propagating from one cell to the next, and Arum forms, which have hyphae developing in homes between plant cells [42]. Although some families or species have both types, the decision between Paris and Arum is largely influenced by the host family [42]. Host plants affect ERH proliferation and arbuscules formation [1]. Plant chromatin is depolymerized from body material, which indicates increased transcription of plant deoxyribonucleic acid (DNA) in arbuscules cells [42]. Major alterations are needed within the plant host cell to accommodate the arbuscules. The vacuoles contract and various cellular organelles proliferate. The cytoskeleton structure of plant cells surrounds the arbuscules organisation. There are two different types of hyphae that come from the roots of the host plants being colonised: after colonisation occurs, transient runner hyphae grow from the roots of the plants to the ground soil. These are ERHs that absorb phosphorus and other nutrients into plants. The hyphae of AM fungi have a high quantitative surface area to volume quantitative ratio, which means that their absorption capacity is greater than that of plant roots [43]. The hyphae of AM fungi are also smaller than roots and can penetrate into soil pores where roots cannot enter [1, 44]. The fourth type of hyphae of AM fungi is different in morphology, it grows from roots and colonises different roots of host plants [40].

4.4 Multiplicity of AM fungi and dominant genera

There are 336 species of AM fungi. Among them, the dominant genera include 6 species, including Acaulospora, Glomus, Gigaspora, Scutellospora and Entrophospora, which have greater advantages in farmland than uncultivated ones (on Google.com). Glomus is the dominant genus, which can be obtained on land all over the world and reproduced by biostimulants.

4.5 AM fungi characteristics and utilisation

The symbiotic relationship of AM fungi is a traditional instance of a mutualistic relationship that can regulate plant growth and development. The fungal mycelium network extends under the roots of the plant, facilitating the absorption of nutrients uptake (NU) that are otherwise unavailable. The mycelium of AM fungi colonises the roots of many different plant species, forming a common mycorrhizal network (CMN). Common mycorrhizal network is considered to be the main component of the terrestrial ecosystem (TES) and has a profound impact on various plant communities, especially on invasive plants [1, 15, 20, 45], and the fungal removal of phosphorus and nitrogen (N) are transferred to plants [6, 31, 46, 47]. In addition, the transfer of common nutrients from fungi to plants has a variety of side effects and improves plant resistance to biological and non-biological factors. They have the ability to improve soil properties, thereby stimulating plant improvement under normal conditions and under stress [47, 48]. The colonisation of AM fungi increases the plant's resistance to stressful signals, which leads to its morphological and physiological characteristics having a large number of changes [48, 49]. AM fungi are considered to be a natural growth regulator for most terrestrial plants. AM fungi are used as biological vaccines (bioinoculants), and researchers are promoting their use as excellent biological fertilisers to achieve sustainable crop yields. Constant mass and significantly higher extraradical hyphae mycelium [1, 4, 15, 20, 50]. Glomalin-related soil protein (GRSP) is believed to maintain the water content of soil exposed to various abiotic stresses [51], then adjust the water frequency between soil and plants and automatically trigger plant improvement. Glomalin contains 30–40% carbon and its related compounds, which can prevent soil from drying out by increasing its water holding capacity [52]. Growth associated functions, including stomatal conductance, leaf water potential (LWP), relative water content (RWC), PSII efficiency and CO_2 assimilation, depend on AM fungal inoculation [15, 53]. AM fungi also help increase resistance to water stress through physiological changes in organs and tissues on the earth [54]. AM fungi can improve the accumulation of dry matter and improve the absorption of water, thereby enhancing the plant's resistance to stress. The use of AM fungi for plant growth in various biological [55] ecosystems can make a significant contribution to the cultivation of organic culturing to stimulate growth and increase yield.

5. AM fungi for environmental implication

AM fungi are extremely beneficial to the environment and make a significant contribution to improving soil and plant health and maximising the intake of macro and micronutrients. This symbiotic relationship between fungi and plants spans millions of years, and these characteristics allow plants to survive. Colonise areas that are difficult to resist; however, their presence in the soil makes them vulnerable to erosion and tilling. Tilling reduces the effectiveness of soil inoculation and fungi by destroying the mycelial network.

6. AM fungi utilisation as a biofertilizer and substitute the chemical fertiliser

AM fungus produces glomalin protein in the soil environment, which may promote soil particle aggregation. It also boosts soil oxygen and carbon content, which is beneficial to plant and soil health. AM fungal-mediated plant growth is accelerated by a factor of 10, allowing for faster plant establishment. It improves standard root biomass and root yield in cereals, legumes, vegetables, spices, and fruits crops by up to 50 times. AM fungus inoculums are a mixture of naturally occurring material (spores, root bit, hypha, mycelium and substrate) used to improve soil fertility, production and importance in agroecosystems. AM fungi, like plant growth and development, are extremely important to soil health (SH). Various studies and research on AM fungus have been conducted over the last three decades, highlighting its numerous benefits to soil and plant health as well as crop productivity (CP). As a result, it is widely assumed that AM fungi might be considered as a chemical fertiliser (CF) substitute due to the fact that the utility of AM fungi can effectively minimise the quantitative usage of chemical fertilisers input [1, 4, 15]. Through their poor impact on the quality of food products, soil health, air, and water systems, the continued use of lifeless chemical fertilisers, herbicides and fungicides has caused a slew of problems for soil, plants and human health (HH) [47, 56]. It is estimated that AM fungus can reduce the use of chemical fertilisers by up to 50% for pleasant agricultural output; however, this estimate is dependent on plant species morphology and traditional traumatising regimes.

7. AM fungi nutrients translocation and exchange effectiveness

AM fungi have a mutually beneficial symbiotic relationship with the host. These biologically active phytochemicals and AM fungi participate in the interaction between plants and soil microorganisms. They have limited saprobic ability and rely on host plants as their carbon nutrient for food. The photosynthetic product of the host plant in the form of hexose. The transfer of carbon from plants to fungi can also occur through arbuscules or intraradical hyphae [1, 33, 57]. The intraradical mycelium is where AM fungus perform secondary hexose production (IRM). Hexose is metabolised to trehalose and glycogen in the mycelium. Trehalose and glycogen are carbon storage forms that can be swiftly generated and degraded, and they can help to buffer intracellular sugar levels [4]. The intraradical hexose is converted to pentose for nucleic acids via the oxidative pentose phosphate pathway. Lipid production takes place within the intraradical mycelium as well. After that, lipids are stored or exported to extraradical hyphae, where they will be stored or metabolised. Gluconeogenesis is the degradation of lipids into hexoses that occurs in extraradical hyphae [57]. The extraradical hyphae store around a quarter of the carbon transferred from the plant to the fungi [58]. The AM fungus may absorb over 20% of the carbon from the host plant [57]. This reflects the host plant's significant carbon investment in the mycorrhizal network (MNW) and contribution to the organic carbon pool below ground (OCP). AM fungus is escalating uptake and switching of P and exclusive macro and micronutrients from the host plant, increasing the plant's carbon delivery to the AM fungi. Similarly, nutrient uptake and transfer are reduced, as is the amount of photosynthate available to the fungi. The ability of different AM fungus species to supply nutrients to the plant varies. AM fungi can be poor symbionts in some situations, delivering little P while using large amounts of carbon [59]. The primary benefit of AM fungus to plants has been related to their ability to absorb nutrients/vitamins over longer periods of time, particularly P. AM fungus may be far more efficient at absorbing P than plant roots. Diffusion transports phosphorus and other minerals to the roots, and hyphae shorten the distance necessary for diffusion, resulting in improved uptake. The rate of phosphorus deposition in AM fungus could be six times that of root hairs [44]. In some situations, the mycorrhizal network can totally take over the role of phosphorus and nutrient absorption, and all of the plant's phosphorus can come from hyphal sources [59]. Although mycorrhizas have been discovered in watery situations, wet soils have been demonstrated to impair mycorrhizal colonisation in numerous species [60].

8. Role of AM fungi in mineral nutrition and their impact on symbiotic host

As many reports have emphasised, overexploitation of land usually has serious consequences for biodiversity, which in turn will have additional impacts on ecosystem functions. AM fungi are very beneficial to increase nutrient bioavailability, which can reduce irrigation and increase fertilisation efficiency. In this symbiotic relationship, an important role is to transport nutrients from organic carbon (OC) in the form of lipids and sugars [61]. It is believed that mycorrhizal colonisation stimulates the absorption of nutrients by plants. This leads to accelerated production of photosynthesis, thereby accelerating biomass accumulation [4, 12, 46, 62]. AM fungi can improve the absorption of inorganic nutrients by almost all plants, especially phosphorus [1, 6, 12, 46]. AM fungi are also very effective in helping plants absorb nutrients from nutrient-poor soils. In addition to macronutrients, AM fungi have been reported to increase the plant utilisation of micronutrients such as zinc, iron and copper [4, 15]. AM fungi increase the absorptive capacity of the host root surface. Experimental results on tomato plants inoculated with AM fungi showed that the leaf area and N, K, calcium and P content increased, indicating that the plant is growing well [63]. AM fungi coexist with roots to obtain important nutrients from the host plant, thereby providing mineral nutrients such as N, P, K, Ca, Zn and S. Therefore, AM fungi can support plant vegetative root cells even when it is not important. AM fungi produce arbuscules fungal structures (AFS), which promote the exchange of inorganic minerals and C and P compounds, and ultimately transfer large amounts of energy to the host plant [4, 64, 65]. AM fungi have been found to help the absorption of P and N, and ultimately help improve plants in better areas and reduce P.

Under drought stress, the symbiosis of AM fungi undoubtedly increases the concentration of N, P and Fe in *Pelargonium graveolens* L. [66]. Gomez Bellot et al. [67] believed that AM fungi can increase the absorption of almost all essential nutrients, on the contrary, can reduce the absorption of Na and Cl, thereby stimulating growth [68, 69]. Many scientists have discovered that AM fungi play a significant role in absorbing nutrients from the soil, especially N and P can effectively promote the growth of host plants. Many studies have shown that AM fungi have the ability to absorb N and transfer it to nearby plants or hosts. The symbiosis of AM fungi produces huge underground extraradical mycelium from the roots and surrounding rhizosphere, which helps to improve nutrient absorption. In the case of increased environmental concentration and CO_2 content [69], AM fungi are said to promote the growth and accumulation of micro and macronutrients and their distribution in seedlings that grow with an accelerated increase in manganese range [6, 63, 70]. Improving plant nutrition and maintaining the ratio of Ca^{2+} and Na^{+} are essential dynamic properties that help enhance AM fungal colonisation of beneficial ingredients in multifunctional plant performance. Improved growth and levels of protein, Fe and Zn had been discovered in mycorrhizal chickpea [71]. In addition, various reports have shown that the mycorrhiza of *Lotus japonicus* root has excellent K^{+} transporter activity [72]. In addition, the meta-analysis report confirmed the symbiosis effect of mycorrhiza and a variety of micronutrients in crops. Multiple inspections of the previous pair at the same stage. Over the years, it has been shown that AM fungi (*Glomus mosseae* and *Rhizophagus irregularis*) increase the translocation of heavy metals within the shoots [73].

9. Role of AM fungi in plant productivity and quality

AM fungi are no longer the most effective, they can increase the nutritional value of plants, and they can also increase the quality and quantity of plants. Improve the nutritional quality of plants through exposure and production of carotenoids and some volatile compounds [74, 75]. Prasad [63] found that AM fungi having a positive effect on the quality and yield of nightshade solanaceous crops (tomatoes, potatoes and eggplants). Zeng et al. [76] mentioned modified sugars, organic acids, vitamin C, flavonoids and minerals from *Glomus versiforme* to produce better quality citrus fruits. The symbiotic relationship of AM fungi can induce a more adequate accumulation of anthocyanins, chlorophylls, carotenoids, overall soluble phenols, tocopherols and many minerals [77–79]. AM fungi have been used in large scale field production of corn [80], yams [81], potatoes [82], soybeans [83–85] and onions [6], confirming that AM fungi have a significant increase in production. AM fungi can also promote the biosynthesis of valuable phytochemicals in edible plants and lead them into the healthy food chain [86].

10. Role of AM fungi in enhance production of growth hormones for host

Plants with AM fungi have higher levels of growth regulators, such as cytokinins and auxins than those without mycorrhiza. AM fungi colonised roots display adjustments in root morphology through acquiring plentiful thicker and delivering fewer root hairs. Host tissue is affected by mycorrhizal colonisation. It is suitable for cytokinin, abscisic acid and gibberellin-like substances. The influence of AM fungi on photosynthesis and host morphology can also be hormones.

11. Role of AM fungi in abiotic stresses

11.1 AM fungi drought tolerance activity

Plants inoculated with AM fungi are tolerant of drought, because these AM fungi help absorb toxic minerals and improve the overall health of plants and soil, toxic levels and mineral toxicity [4, 6, 15]. AM fungi help plants to absorb nutrients from the soil in exchange for sugar produced by the plants. In the forest ecosystems, ectomycorrhizas form filaments called hyphae net, which run between trees to act as connecting bonds. This huge underground transportation network is called the common mycorrhizal network. a common mycorrhizal network uses chemical communication to exchange nutrients between trees when needed. A common mycorrhizal network also makes it easier for trees to obtain water that cannot be reached by their roots. In the presence of excessive soil temperature, soil toxins, and extreme soil pH, plants treated with AM fungi can improve drought tolerance and survival.

11.2 AM fungi salinity tolerance effectiveness

Salt stress is believed to inhibit plant growth through use, affect nutrient improvement and net assimilation rate, resulting in a decline in productivity. It also contributed to the beginning of the era of excessive reactive oxygen species [87, 88]. Soil contaminated by salt and the correct use of AM fungi to reduce salt content have harmful effects on plants [89]. Several studies have shown that AM fungi improve plant growth and productivity under salt stress conditions [90]. AM fungus improved the growth rate, leaf water potential (LWP), and water usage efficiency (WUE) of Antirrhinum majus plants, according to El-Nashar [91]. Under salinity, Ait-El-Mokhtar et al. [92] found that the AM fungal symbiosis improved physiological parameters, photosynthetic rate, stomatal conductance and leaf water relations. Under saltwater circumstances, AM fungus inoculated on Allium plants showed better development, including leaf area index, fresh and dried biomass [6, 93]. Under salt stress conditions, the concentrations of total P, Ca^{2+}, N, Mg^{2+} and K^{+} in cucumber plants treated with AM fungi are higher than those of uninoculated plants [94]. Pepper exhibits better chlorophyll content and better Mg^{2+} and N absorption, while at the same time reducing Na^{+} transmission under salt conditions [95]. Inoculation with AM fungi can effectively regulate the level of major growth regulators. Plants colonised by AM fungi can reduce oxidative stress by inhibiting lipid membrane peroxidation under salt stress conditions [90, 96].

11.3 AM fungi heavy metals tolerance activity

It is generally believed that AM fungi can promote the rooting of plants in heavy metal contaminated soil because they can improve the plant defence system mediated by AM fungi and promote their growth and expansion. Heavy metals can also be obtained from food crops, fruits and vegetables and soils, inflicting numerous health hazards. The association between AM fungi and wheat actively increases nutrient uptake under aluminium stress [97]. Heavy metals can be fixed in the hyphae of endogenous and exogenous fungi [98], they have the ability to fix heavy metals in the cell wall and accumulate in vacuoles or can also chelate with some other substances in the cytoplasm [99], and then reduce the toxicity of metals in plants. The more common reason is that these fungi can improve the morphological and physiological processes of the rapid evolution of plant biomass, thereby promoting the absorption of fixed essential nutrients (copper, zinc, phosphorus, nitrogen, potassium). The toxicity in the host organism can be reduced by AM fungal mediated plants [15, 20, 100]. It is likewise believed that improved growth or chelation in the rhizospheric soil can cause metal dilution in plant tissues [101]. It has been reported that AM fungi bind to Cd and Zn in the cell wall and cortical cells of the mantle hyphae, limiting their absorption and leading to better growth, yield and nutritional status [20]. It has high cation exchange and metal absorption potential [102]. Similarly, AM fungi can also solve the problem of low Cd mobility and toxicity by increasing the pH of the soil [103], reducing Cd to extraradical mycelium [104], and combining Cd with the glomalin, a glycoprotein. AM fungi are very effective in reducing the level of Cd in each vacuole sand cell wall, which roughly contributes to the detoxification of Cd in rice [105].

11.4 Role of AM fungi in high and low temperature tolerance to crops

As soil temperature increases, the response of the plant community may depend on the interaction of AM fungi to ensure sustainable production. Biomass production, withering and burning of leaves and reproductive organs, leaf tearing and ageing, additional damage leading to fruit discolouration, reduced yield and cell death, and related oxidative stress increases. In general, plants treated with AM fungi perform better under heat stress. Maya and Matsubara [106] pointed out the relationship between *Glomus fasciculatum* and plant growth and the most important positive growth changes under high-temperature conditions. AM fungi can increase the resistance of plants to low temperatures. In addition, most information claims that some plants inoculated with low-temperature AM fungi grow and spread better than plants not inoculated with AM fungi [107, 108]. AM fungi help plants prevent cold stress and ultimately accelerate plant development [108]. AM fungi can also maintain water in the host plant and increase phytochemicals, thereby strengthening the plant's immune system and increasing protein levels to help the plant fight cold stress [109]. The symbiosis of AM fungi improves the relationship between water and plants, increasing the possibility of gas exchange and osmotic regulation [110]. AM fungi increase the synthesis of chlorophyll leading to a noteworthy perfection in the concentrations of numerous metabolites in plants subjected to cold stress conditions [109, 111].

12. Role of AM fungi in seed production, offspring fertility and tolerances to disease and pest

AM fungi can improve the vigour of offspring, and through AM fungi, the fertility and survival rate of plant seeds can be improved. AM fungi increase the resistance of pests to root and soil-borne pathogens through mycorrhizal induced resistance and the production of secondary metabolites and increase resistance to leaf pathogens [1].

13. Role of AM fungi in carbon cycling and phytoremediation

The production of glycoproteins, including glomalin that may be involved in the formation and stability of soil aggregates, should have an additional impact on the unusual microorganisms associated with AM fungal mycelium [41, 111]. Changes in native plant groups in areas threatened by desertification are often related to the deterioration of soil physical and biological properties, soil structure, availability of nutrients and organic matter and physical properties of soil [4]. A particularly new method of land restoration is to inoculate the soil with AM fungi and at the same time reintroduce vegetation to ecological reclamation. This allows host plants to take root in degraded soil and improves soil quality and health [1]. In the long run, compared with unmodified soil and soil inoculated with single exotic species of AM fungi, the introduction of a mixture of natural AM fungi resulted in significantly improved soil quality parameters [4]. The advantages included increased plant growth, increased P absorption [4] and soil N content, higher soil organic matter, soil aggregation, which was linked to stronger legume nodulation in the presence of AM fungus, higher water infiltration, and soil aeration. The native AM fungus aids in the removal of heavy metals from contaminated soil, making it healthier and more conducive to agricultural development [4].

14. Role of AM fungi on global climate change

Climate change poses a major threat to AM fungi due to irreparable damage to various ecosystems, in addition to increasing habitat loss because of human activities, so gradual steps must be taken to mitigate the next errors that occur from these concerns. Global climate change is affecting the population of AM fungi and the interaction between AM fungi and their host plants. It is generally believed that the interaction between organisms can affect their response to global climate change. In a recent meta-analysis, it was found that AM fungi increased plant biomass under drought conditions. The AM fungus itself has been shown to increase its biomass in response to accelerated emissions of carbon dioxide into the atmosphere. Climate change has brought challenges to the supply of water, food, and nutrients. There may also be deficits in some places, and surpluses in other places. The relationship between forest trees and AM fungi helps them, mainly based on the percentage of sources needed, and may help us solve these problems.

15. Conclusion

Several studies have recognised the positive role of AM fungi in improving plant growth in stressful environments, hence, this manuscript consistently combines existing evidence about the AM fungi general reproduction, lifestyle, and its widespread distribution to generate knowledge for agriculturists and researchers. The symbiosis and courtship of AM fungi and different plants in a stressful environment. AM fungi contribute to many aspects of plant life, especially better nutrition, better growth, stress and disease resistance. The particles accumulate to improve the soil's resistance to wind and water erosion. AM fungi reduce the leaching of nutrients in the soil, thereby promoting the retention of nutrients in the soil and reducing the risk of groundwater pollution. These multiple benefits of AM fungi translate into an important ecological benefit in natural circumstances. Formerly, AM fungi had been particulars mentioned as useful entities for nutrient uptake from soil; however, it has recently been honestly described that plants inoculated with AM fungi can effectively control various environmental factors such as salt, drought, nutrient stress, alkali stress, cold stress and high temperature, thereby controlling plant yield and productivity. A wide range of plants such as beans, oilseeds, fruits, fibre plants, vegetables, forests, and nurseries are developed. Promoting the use of additively manufactured AM fungi is critical to the sustainability of modern global agricultural systems. Agricultural development can significantly reduce the use of artificial (lifeless chemical fertilisers, pesticide, insecticide) fertilisers and various chemicals, thereby supporting biologically healthy agriculture. Inoculation with AM fungi that can increase plant growth and productivity can help meet the consumer needs of growing populations around the world. In addition, environmental protection technology can only be protected through widespread usage. The most important knowledge in fortune research should be to control AM fungal inoculum mediated growth and improve the identity of genes and gene products downregulated by stress signals. AM fungi regulate tolerance mechanisms, in addition, activating crosstalk to control the overall performance of the plant can help increase crop yields. AM fungi must be explored in any respect ranges to extra inspect their role in the landscape as a biofertilizer for sustainable agricultural production for fast increase population worldwide.

Conflict of interest statements

There is no conflict of interest.

Author details

Kamal Prasad*, Agam Khare and Prateek Rawat
Absolute Foods, Division of Microbiology, AgriScience, Gurugram, Haryana, India

*Address all correspondence to: kamalprsd27@gmail.com

References

[1] Prasad K. Studies on Ecological Factors Affecting Vesicular Arbuscular Mycorrhizal Infection in Sugarcane. Muzaffarpur, Bihar, India: B.R. Ambedkar Bihar University; 1993

[2] Smith SE, Read DJ. Mycorrhizal symbiosis. San Diego, London, New York, Boston, Sydney, Tokyo, Toronto: Academic Press, Harcourt Brace and Company, Publishers; 1997

[3] Prasad K. Biofertilizers: A new dimension for agriculture and environmental development to improve production in sustainable manner. Journal of Basic and Applied Mycology. 2015;**11**(1& II):5-13

[4] Prasad K. Biology, diversity and promising role of mycorrhizal entophytes for green technology. In: Maheshwari DK, editor. Endophytes: Biology and Biotechnology, Series Sustainable Development and Biodiversity 15. Springer International Publishing AG: Switzerland; 2017. pp. 257-301

[5] Bisleski RL. Phosphate pools, phosphate transport, and phosphate availability. Annual Review of Plant Physiology. 1973;**24**:225-252

[6] Prasad K. Impact of biological fertilizer arbuscular mycorrhizal fungi and conventional fertilizers mobilization on growth, yield, nutrient's uptake, quercetin and Allin contents in allium crops cultivation under field conditions in semi-arid region of India. South Asia Journal of Experimental Biology. 2021;**11**(1):15-26

[7] Janos DP. Mycorrhizae influence tropical succession. Biotropica. 1980;**12**:56-64

[8] Allen EB, Allen MF. Water relations of xeric grasses in the field: Interactions of mycorrhizae and competition. New Phytologist. 1986;**104**:559-571

[9] Fitter AH. Influence of mycorrhizal infection on competition for phosphorus and potassium by two grasses. New Phytologist. 1977; **79**:119-125

[10] Newman EI. Mycorrhizal links between plants: Their functioning and ecological significance. Advances in Ecological Research. 1988;**18**:243-270

[11] Koske RE, Sutton JC, Sheppard BR. Ecology of Endogone in Lake Huron sand dunes. Canadian Journal of Botany. 1975;**53**:87-93

[12] Prasad K. Diversification of Glomermycota form arbuscular mycorrhizal fungi associated with vegetable crops cultivated underneath natural ecosystems in arid region of Rajasthan, India. Current Investigations in Agriculture and Current Research. 2021;**9**(2):1205-1212

[13] Dehn B, Schuepp H. Influence of VA mycorrhizae on the uptake and distribution of heavy metals in plants. Agriculture and Environment. 1989;**29**:79-83

[14] Prasad K, Rajak RC. Biotechnological application of mycorrhizae in reclamation of mined dumps. In: Bagyaraj DJ, Tilak KVBR, Kehri HK, editors. Integrated Management of Plant Resources. Jodhpur: Scientific Publishers (India); 2000. pp. 283-292

[15] Prasad K. Arbuscular mycorrhizal fungi and plant collaborations influences ecology and environmental changes for global sustainable development. Journal of Ecology and Natural Resources. 2021;**5**(1):1-16

[16] Stahl PO, Smith WK. Effects of different geographic isolates of *Glomus* on the water relations of *Agropyron smithii*. Mycologia. 1984;**76**:261-267

[17] Schonbeck F. Einfluss der endotrophen Mykorrhiza auf die Krankheitsresistenz hoherer Pflanzen. Z. PflKrank. PflSchutz. 1978;**85**: 191-196

[18] Prasad K, Warke RV, Khadke K. Management of soilborne pathogens to improve production of pulses using organic technologies for sustainable agriculture. International Journal of Research and Analytical Reviews. 2019;**6**(2):82-101

[19] Graham JH. Assessing costs of arbuscular mycorrhizal symbiosis in agro ecosystems. In: Podila GK, Douds DD Jr, editors. Current Advances in Mycorrhizal Research. St. Paul, MN: APS Press; 2000. pp. 127-140

[20] Prasad K. Positive importance of arbuscular mycorrhizal fungi for global sustainable agriculture and environment management for green technology. Current Investigations in Agriculture and Current Research. 2020;**9**(2): 1182-1185

[21] Mosse B. Plant growth response to vesicular-arbuscular mycorrhizae. X. Response of Stylosanthes and maize to inoculation in unsterile soils. New Phytologist. 1973;**78**:277-288

[22] Morton JB. Problems and solutions for the integration of glomalin taxonomy, systematic biology, and the study of endomycorrhizal phenomena. Mycorrhiza. 1993;**2**:97-109

[23] Morton JB. Evolution of endophytism in arbuscular mycorrhizal fungi of Glomales. In: Bacon CW, White JH, editors. Microbial Endophytes. New York: Marcel Dekker Inc; 2000. pp. 121-140

[24] Mosse B. Vesicular-Arbuscular Mycorrhiza Research for Tropical Agriculture. Hawaii Institute of Tropical Agriculture and Human resources: University of Hawaii; 1981

[25] Morton JB, Redecker D. Two families of Glomales, Archaeosporaceae and Paraglomaceae, with two new genera Archaeospora and Paraglomus, based on concordant molecular and morphological characters. Mycologia. 2001;**93**:181-195

[26] Sieverding E, Oehl F. Revision of *Entrophospora* and description of *Kuklospora* and *Intraspora*, two new genera in the arbuscular mycorrhizal Glomeromycetes. Journal of Applied Botany and Food Quality. 2006;**80**:69-81

[27] Spatafora JW, Chang Y, Benny GL, Lazarus K, Smith ME, Berbee ML. A phylum-level phylogenetic classification of zygomycete fungi based on genome-scale data. Mycologia. 2016;**108**: 1028-1046

[28] Redecker D, Schussler A, Stockinger H, Sturmer SL, Morton JB, Walker C. An evidence-based consensus for the classification of arbuscular mycorrhizal fungi (Glomeromycota). Mycorrhiza. 2013;**23**(7):515-531

[29] Bago B, Pfeffer PE, Shachar-Hill Y. Carbon metabolism and transport in arbuscular mycorrhizas. Plant Physiology. 2000;**124**:949-958

[30] Jiang YN, Wang WX, Xie QJ, Liu N, Liu LX, Wang DP. Plants transfer lipids to sustain colonization by mutualistic mycorrhizal and parasitic fungi. Science. 2017;**356**:1172-1175

[31] Smith S, Read D. Mycorrhiza symbiosis. 3rd ed. San Diego, CA: Academic Press; 2008

[32] Kar RK, Mandaokar BD, Kar R. Mycorrhizal fossil fungi from the Miocene sediments of Mirozam, Northeast India. Current Science. 2005;**89**:257-259

[33] Remy W, Taylor T, Hass H, Kerp H. Four hundred-million-year-old vesicular arbuscular mycorrhizae.

Proceedings of the National Academy of Sciences of the United States of America. 1994;**91**(25):11841-11843

[34] Wang B, Yeun LH, Xue Y, Liu Y, Ane JM, Qiu YL. Presence of three mycorrhizal genes in the common ancestor of land plants suggests a key role of mycorrhizas in the colonization of land by plants. New Phytologist. 2010;**186**(2):514-525

[35] Nagahashi G, Douds DD, Abney GD. Phosphorus amendment inhibits hyphal branching of VAM fungus Gigaspora margarita directly and indirectly through its effect on root exudation. Mycorrhiza. 1996;**6**(5): 403-408

[36] Douds DD, Nagahashi G. Signalling and recognition events prior to colonisation of roots by arbuscular mycorrhizal fungi. In: Podila GK, Douds DD, editors. Current Advances in Mycorrhizae Research. Minnesota: APS Press; 2000. pp. 11-18

[37] Sbrana C, Giovannetti M. Chemotropism in the arbuscular mycorrhizal fungus glomus mosseae. Mycorrhiza. 2005;**15**(7):539-545

[38] Parniske M. Arbuscular mycorrhiza: The mother of plant root endosymbiosis. Nature Reviews Microbiology. 2008;**6**:763-775

[39] Gianinazzi S, Gianinazzi-Pearson V. Progress and headaches in endomycorrhiza biotechnology. Symbiosis. 1986;**2**:139-149

[40] Wright SF. Management of arbuscular mycorrhizal fungi. In: Zobel RW, Wright SF, editors. Roots and Soil Management: Interactions between Roots and the Soil. USA: American Society of Agronomy; 2005. pp. 183-197

[41] Prasad K, Warke RV. Biology, diversity and promising role of microbial inoculations specially mycorrhizae in agriculture and natural ecosystems to improve productivity for sustainable development and environmental sustainability. In: National Conference, GKV, Hardwar, India. 2018. pp. 7-8

[42] Armstrong L, Peterson RL, Armstrong L, Peterson RL. the interface between the arbuscular mycorrhizal fungus *Glomus intraradices* and root cells of *Panax quinquefolius*: A paris-type mycorrhizal association. Mycologia. 2002;**94**(4):587-595

[43] Turrini A, Bedini A, Loor MB, Santini G, Sbrana C, Giovannetti M. Local diversity of native arbuscular mycorrhizal symbionts differentially affects growth and nutrition of three crop plant species. Biology and Fertility of Soils. 2018;**54**:203-217

[44] Bolan NS. A critical review of the role of mycorrhizal fungi in the uptake of phosphorus by plants. Plant and Soil. 1991;**134**(2):189-207

[45] Pringle A, Bever JD, Gardes M, Parrent JL, Rillig MC, Klironomos JN. Mycorrhizal symbioses and plant invasions. Annual Review of Ecology, Evolution, and Systematics. 2009;**40**:699-715

[46] Prasad K. Effect of dual inoculation of arbuscular mycorrhiza fungus and cultivar specific *Bradyrhizobium Japonicum* on the growth, yield, chlorophyll, nitrogen and phosphorus contents of soybean (*Glycine max* (L.) Merrill.) grown on alluvial soil. Journal of Innovation in Applied Research. 2021;**4**(1):1-12

[47] Prasad K. Advantages and nutritional importance of organic agriculture produces food on human, soil, and environmental health in modern lifestyle for sustainable development. Aditum Journal of Clinical and Biomedical Research. 2021;**5**(2):1-7

[48] Alqarawi AA, Abd-Allah EF, Hashem A. Alleviation of salt-induced adverse impact via mycorrhizal fungi in Ephedra aphylla Forssk. Journal of Plant Interactions. 2014;**9**(1):802-810

[49] Hashem A, Abd_Allah EF, Alqarawi AA, Aldubise A, Egamberdieva D. Arbuscular mycorrhizal fungi enhance salinity tolerance of *Panicum turgidum Forssk* by altering photosynthetic and antioxidant pathways. Journal of Plant Interactions. 2015;**10**(1):230-242

[50] Syamsiyah J, Herawati A, Mujiyo. The potential of arbuscular mycorrhizal fungi application on aggregate stability in Alfisol soil. IOP Conference Series: Earth Environmental Science. 2018;**142**:012045

[51] Wu Z, McGrouther K, Huang J, Wu P, Wu W, Wang H. Decomposition and the contribution of glomalin-related soil protein (GRSP) in heavy metal sequestration: Field experiment. Soil Biology and Biochemistry. 2014;**68**:283-290

[52] Sharma S, Prasad R, Varma A, Sharma AK. Glycoprotein associated *with Funneliformis coronatum, Gigaspora margarita* and *Acaulosporascrobiculata* suppress the plant pathogens in vitro. Asian Journal of Plant Pathology. 2017;**11**(4):192-202

[53] He F, Sheng M, Tang M. Effects of Rhizophagus irregularis on photosynthesis and antioxidative enzymatic system in *Robinia pseudoacacia* L. under drought stress. Frontiers in Plant Science. 2017;**8**:183-186

[54] Barzana G, Aroca R, Paz JA, Chaumont F, Martinez-Ballesta MC, Carvajal M. Arbuscular mycorrhizal symbiosis increases relative apoplastic water flow in roots of the host plant under both well-watered and drought stress conditions. Annals of Botany. 2012;**109**:1009-1017

[55] Liu C, Ravnskov S, Liu F, Rubaek GH, Andersen MN. Arbuscular mycorrhizal fungi alleviate abiotic stresses in potato plants caused by low phosphorus and deficit irrigation/ partial root-zone drying. Journal of Agricultural Science. 2018;**156**:46-58

[56] Yang S, Li F, Malhi SS, Wang P, Dongrang S, Wang J. Long term fertilization effects on crop yield and nitrate nitrogen accumulation in soil in North-Western China. Agronomy Journal. 2004;**96**:1039-1049

[57] Pfeffer P, Douds D, Becard G, Shachar-Hill Y. Carbon uptake and the metabolism and transport of lipids in an arbuscular mycorrhiza. Plant Physiology. 1999;**120**(2):587-598

[58] Hamel C. Impact of arbuscular mycorrhiza fungi on N and P cycling in the root zone. Canadian Journal of Soil Science. 2004;**84**(4):383-395

[59] Smith S, Smith A, Jakobsen I. Mycorrhizal fungi can dominate phosphate supply to plants irrespective of growth responses. Plant Physiology. 2003;**133**(1):16-20

[60] Smith SE, Read DJ. Mycorrhizal Symbiosis. London: Academic Press; 2002

[61] Luginbuehl LH, Menard GN, Kurup S, Van Erp H, Radhakrishnan GV, Breakspear A. Fatty acids in arbuscular mycorrhizal fungi are synthesized by the host plant. Science. 2017;**356**: 1175-1178

[62] Mitra D, Navendra U, Panneerselvam U, Ansuman S, Ganeshamurthy AN, Divya J. Role of mycorrhiza and its associated bacteria on plant growth promotion and nutrient management in sustainable agriculture. International Journal of Life Sciences and Applied Science. 2019;**1**:1-10

[63] Prasad K. Influence of arbuscular mycorrhizal fungal biostimulants ad

conventional fertilizers on some solanaceous crops for growth, productivity and nutrient stoichiometry under field conditions in semi- arid region of Maharashtra, India. Journal of Experimental Biology and Agricultural Sciences. 2021;**9**(1):75-86

[64] Li X, Zeng R, Liao H. Improving crop nutrient efficiency through root architecture modifications. Journal of Integrative Plant Biology. 2016;**58**:193-202

[65] Liu LZ, Gong ZQ, Zhang YL, Li PJ. Growth, cadmium uptake and accumulation of maize *Zea mays* L. under the effects of arbuscular mycorrhizal fungi. Ecotoxicology. 2014;**23**:1979-1986

[66] Amiri R, Ali N, Nematollah E, Mohammad RS. Nutritional status, essential oil changes and water-use efficiency of rose geranium in response to arbuscular mycorrhizal fungi and water deficiency stress. Symbiosis. 2017;**73**:15-25

[67] Gomez-Bellot MJ, Ortuno MF, Nortes PA, Vicente-Sanchez J, Banon S, Sanchez Blanco MJ. Mycorrhizal euonymus plants and reclaimed water: Biomass, water status and nutritional responses. Scientia Horticulturae. 2015;**186**:61-69

[68] Evelin H, Giri B, Kapoor R. Contribution of *Glomus intraradices* inoculation to nutrient acquisition and mitigation of ionic imbalance in NaCl-stressed *Trigonella foenum-graecum*. Mycorrhiza. 2012;**22**:203-217

[69] Zhu XC, Song FB, Liu SQ, Liu FL. Arbuscular mycorrhiza improves growth, nitrogen uptake, and nitrogen use efficiency in wheat grown under elevated CO_2. Mycorrhiza. 2016;**26**:133-140

[70] Bati CB, Santilli E, Lombardo L. Effect of arbuscular mycorrhizal fungi on growth and on micronutrient and macronutrient uptake and allocation in olive plantlets growing under high total Mn levels. Mycorrhiza. 2015;**25**(2): 97-108

[71] Pellegrino E, Bedini S. Enhancing ecosystem services in sustainable agriculture: Biofertilization and biofortification of chickpea (*Cicer arietinum* L.) by arbuscular mycorrhizal fungi. Soil Biology and Biochemistry. 2014;**68**:429-439

[72] Battini F, Gronlund M, Agnolucci M, Giovannetti M, Jakobsen I. Facilitation of phosphorus uptake in maize plants by mycorrhizosphere bacteria. Scientific Reports. 2017;**7**:4686-4689

[73] Ali N, Masood S, Mukhtar T, Kamran MA, Rafique M, Munis MFH. Differential effects of cadmium and chromium on growth, photosynthetic activity, and metal uptake of *Linum usitatissimum* in association with *Glomus intraradices*. Environmental Monitoring and Assessment. 2015;**187**(6):311-314

[74] Castellanos-Morales V, Villegas J, Wendelin S, Vierheiling H, Eder R, Cardenas-Navarro R. Root colonization by the arbuscular mycorrhizal fungus *Glomus intraradices* alters the quality of strawberry fruit (*Fragaria ananassa* Duch.) at different nitrogen levels. Journal of the Science of Food and Agriculture. 2010;**90**:1774-1782

[75] Hart M, Ehret DL, Krumbein A, Leung C, Murch S, Turi C. Inoculation with arbuscular mycorrhizal fungi improves the nutritional value of tomatoes. Mycorrhiza. 2015;**25**:359-376

[76] Zeng L, JianFu L, JianFu L, MingYuan W. Effects of arbuscular mycorrhizal (AM) fungi on citrus quality under nature conditions. Southwest China Journal of Agricultural Science. 2014;**27**:2101-2105

[77] Baslam M, Garmendia I, Goicoechea N. Arbuscular mycorrhizal fungi (AMF) improved growth and nutritional quality of greenhouse grown lettuce. Journal of Agricultural and Food Chemistry. 2011;**59**:5504-5515

[78] Prasad K. Improvement of biomass yields of *Terminalia Arjuna* L. through vesicular arbuscular mycorrhizal fungi (*Glomus fasciculatum*) under nursery and field condition. Journal of Phytological Research. 1997; **10**(1-2):39-41

[79] Prasad K. Interaction between *Glomus fasciculatum* AMF and Rhizobium and their effect of *Prosopis juliflora* in nursery conditions. Journal of Basic Applied Mycology. 2002;**1**(1):130-143

[80] Sabia E, Claps S, Morone G, Bruno A, Sepe L, Aleandri R. Field inoculation of arbuscular mycorrhiza on maize (*Zea mays* L.) under low inputs: Preliminary study on quantitative and qualitative aspects. Italian Journal of Agronomy. 2015;**10**:30-33

[81] Lu F, Lee C, Wang C. The influence of arbuscular mycorrhizal fungi inoculation on yam (*Dioscorea* spp.) tuber weights and secondary metabolite content. Peer Journal. 2015;**3**:12-66

[82] Hijri M. Analysis of a large dataset form field mycorrhizal inoculation trials on potato showed highly significant increase in yield. Mycorrhiza. 2016;**2**:209-214

[83] Meghavanshi MK, Prasad K, Harwani D, Mahna SK. Response of soybean cultivars towards inoculation with three Arbuscular Mycorrhizal Fungi (AMF) and *Bradyrhizobium japonicum* in alluvial soil. European Journal of Soil Biology. 2008;**44**:316-323

[84] Meghavanshi MK, Prasad K, Mahna SK. Symbiotic potential, competitiveness and compatibility of indigenous *Bradyrhizobium japonicum* isolates to three soybean genotypes of two distinct agro-climatic regions of Rajasthan, India. Saudi Journal of Biological Sciences. 2010;**17**:303-310

[85] Meghavanshi MK, Prasad K, Mahna SK. Identification of pH tolerant *Bradyrhizobium japonicum* strains and their symbiotic effectiveness in soybean (*Glycine max* (L.) Merr.) in low nutrient soil. African Journal of Biotechnology. 2005;**4**(7):663-666

[86] Rouphael Y, Franken P, Schneider C, Schwarz D, Giovannetti M, Agnolucci M. Arbuscular mycorrhizal fungi act as bio-stimulants in horticultural crops. Scientia Horticulturae. 2015;**196**:91-108

[87] Ahanger MA, Tomar NS, Tittal M, Argal S, Agarwal RM. Plant growth under water/salt stress: ROS production; antioxidants and significance of added potassium under such conditions. Physiology and Molecular Biology of Plants. 2017;**23**(4):731-744

[88] Ahanger MA, Alyemeni MN, Wijaya L, Alamri SA, Alam P, Ashraf M. Potential of exogenously sourced kinetin in protecting Solanum lycopersicum from NaCl-induced oxidative stress through up-regulation of the antioxidant system, ascorbate–glutathione cycle and glyoxalase system. PLoS One. 2018;**13**(9):e0202-e0175

[89] Santander C, Sanhueza M, Olave J, Borie F, Valentine C, Cornejo P. Arbuscular mycorrhizal colonization promotes the tolerance to salt stress in lettuce plants through an efficient modification of ionic balance. Journal of Soil Science and Plant Nutrition. 2019;**19**(2):321-331

[90] Abdel Latef AA, Chaoxing HJ. Does the inoculation with *Glomus mosseae* improve salt tolerance in pepper plants? Plant Growth Regulation. 2014; **33**:644-653

[91] EL-Nashar YI. Response of snapdragon *Antirrhinum majus* L. to blended water irrigation and arbuscular mycorrhizal fungi inoculation: Uptake of minerals and leaf water relations. Photosynthetica. 2017;**55**(2):201-209

[92] Ait-El-Mokhtar M, Laouane RB, Anli M, Boutasknit A, Wahbi S, Meddich A. Use of mycorrhizal fungi in improving tolerance of the date palm (*Phoenix dactylifera* L.) seedlings to salt stress. Science Horticulture. 2019; **253**:429-438

[93] Borde M, Dudhane M, Jite PK. AM fungi influences the photosynthetic activity, growth and antioxidant enzymes in *Allium sativum* L. under salinity condition. Notulae Scientia Biologicae. 2010;**2**:64-71

[94] Hashem A, Alqarawi AA, Radhakrishnan R, Al-Arjani AF, Aldehaish HA, Egamberdieva D. Arbuscular mycorrhizal fungi regulate the oxidative system, hormones and ionic equilibrium to trigger salt stress tolerance in *Cucumis sativus* L. Saudi Journal of Biological Sciences. 2018;**25**(6):1102-1114

[95] Cekic FO, Unyayar S, Ortas I. Effects of arbuscular mycorrhizal inoculation on biochemical parameters in *capsicum annuum* grown under long term salt stress. Turkish Journal of Botany. 2012;**36**:63-72

[96] Talaat NB, Shawky BT. Protective effects of arbuscular mycorrhizal fungi on wheat (*Triticum aestivum* L.) plants exposed to salinity. Environmental and Experimental Botany. 2014;**98**:20-31

[97] Aguilera P, Pablo C, Fernando B, Fritz O. Diversity of arbuscular mycorrhizal fungi associated with *Triticum aestivum* L. plants growing in an and sol with high aluminium level. Agriculture. Ecosystems and Environment. 2014;**186**:178-184

[98] Ouziad F, Hildebrandt U, Schmelzer E, Bothe H. Differential gene expressions in arbuscular mycorrhizal-colonized tomato grown under heavy metal stress. Journal Plant Physiology. 2005;**162**:634-649

[99] Punamiya P, Datta R, Sarkar D, Barber S, Patel M, Da P. Symbiotic role of *Glomus mosseae* in phytoextraction of lead in vetiver grass *Chrysopogon zizanioides* L. Journal of Hazardous Materials. 2010;**177**:465-474

[100] Kanwal S, Bano A, Malik RN. Effects of arbuscular mycorrhizal fungi on metals uptake, physiological and biochemical response *of Medicago sativa* L. with increasing Zn and Cd concentrations in soil. American Journal of Plant Sciences. 2015;**6**:2906-2923

[101] Audet P. Arbuscular mycorrhizal fungi and metal phytoremediation: Ecophysiological complementarity in relation to environmental stress. In: Ahmad P, Rasool S, editors. Emerging Technologies and Management of Crop Stress Tolerance. San Diego: Academic Press; 2014. pp. 133-160

[102] Takacs T, Voros I. Effect of metal non-adapted arbuscular mycorrhizal fungi on Cd, Ni and Zn uptake by ryegrass. Acta Agronomica Hungarica. 2003;**51**:347-354

[103] Shen H, Christie P, Li X. Uptake of zinc, cadmium and phosphorus by arbuscular mycorrhizal maize (*Zea mays L.*) from a low available phosphorus calcareous soil spiked with zinc and cadmium. Environmental Geochemistry and Health. 2006;**28**:111-115

[104] Janouskova M, Pavlikova D. Cadmium immobilization in the rhizosphere of arbuscular mycorrhizal plants by the fungal extraradical mycelium. Plant and Soil. 2010;**332**:511-520

[105] Li H, Luo N, Zhang LJ, Zhao HM, Li YW, Cai QY, et al. Do arbuscular

mycorrhizal fungi affect cadmium uptake kinetics, subcellular distribution and chemical forms in rice? Science of the Total Environment. 2016; **571**:1183-1190

[106] Maya MA, Matsubara Y. Influence of arbuscular mycorrhiza on the growth and antioxidative activity in Cyclamen under heat stress. Mycorrhiza. 2013; **23**(5):381-390

[107] Chen S, Jin W, Liu A, Zhang S, Liu D, Wang F. Arbuscular mycorrhizal fungi (AMF) increase growth and secondary metabolism in cucumber subjected to low temperature stress. Scientia Horticulturae. 2013; **160**:222-229

[108] Birhane E, Sterck F, Fetene M, Bongers F, Kuyper T. Arbuscular mycorrhizal fungi enhance photosynthesis, water use efficiency, and growth of frankincense seedlings under pulsed water availability conditions. Oecologia. 2012;**169**: 895-904

[109] Abdel Latef AA, Chaoxing H. Arbuscular mycorrhizal influence on growth, photosynthetic pigments, osmotic adjustment, and oxidative stress in tomato plants subjected to low temperature stress. Acta physiologiae Plantarum. 2011;**33**:1217-1225

[110] Zhu X, Song F, Liu S, Liu T, Zhou X. Arbuscular mycorrhizae improve photosynthesis and water status of *Zea mays* L. under drought stress. Plant Soil and Environment. 2012;**58**:186-191

[111] Ali N, Masood S, Mukhtar T, Kamran MA, Rafique M, Munis MFH. Differential effects of cadmium and chromium on growth, photosynthetic activity, and metal uptake of *Linum usitatissimum* in association with *Glomus intraradices*. Environmental Monitoring and Assessment. 2015;**187**(6):311-315

Section 3

Fungal Derived Natural Products

Chapter 7

Bioactive Novel Natural Products from Marine Sponge: Associated Fungi

Vasanthabharathi Venkataraman, Kalaiselvi Vaithi and Jayalakshmi Singaram

Abstract

Marine sponges are distributed in the water, from the intertidal zones to thousands of meters deep. They are primitive multicellular invertebrates that live in benthic environments and are bound to solid substrates. Filter feeders, sponges have many microscopic pores on their surface, which allow water to enter and circulate via a network of canals where microbes and organic particles are filtered out and absorbed. Marine fungi are widespread in the oceans and colonize different ecological niches; they are found associated with organisms of all trophic levels and can act as saprobes, symbionts, and parasites. Compared with other marine microorganisms, marine fungus is relatively understudied. Fungi associated with sponges have been discovered to be a promising source of pharmacologically active compounds with unique anticancer, antibacterial, and antiviral properties.

Keywords: sponges, fungus, bioactives, anticancer, antimicrobial

1. Introduction

The ocean is a unique resource that provides a wide range of natural products. The greatest biodiversity is found in ecosystems with high species diversity and population density, such as rocky coasts, kelp beds, and coral reefs [1]. Marine sponges are benthic animals that live in a variety of marine environments. Sponge species diversity is significantly higher in tropical coral reef environments. The ocean is called the "mother of origin of life," and an enormous proportion of all life on Earth exists within the oceans [2].

Marine sponges (Porifera) are the earliest living animal phylum and represent the very beginning of metazoan development. Sponge feeding, as a sedentary organism, sequesters food particles. They can pump and filter a large volume of seawater through a unique and highly vascularized canal system (G [3]).

Sponges are divided into three groups: the *Calcarea* (five orders and 24 families), *Demospongiae* (15 orders and 92 families), and *Hexactinellida* (five orders and 92 families) (six orders and 20 families). About 15,000 sponge species have been identified so far, but their total diversity can be much larger. The 95% of them live in the ocean, with only around 1% existing in freshwater [4].

Sponges are a great place to live not only for macro-organisms such as worms, shrimp, and crabs, but also for a variety of microorganisms, which live in the canals, between cells, and even inside the cells. Large numbers of microorganisms, such as bacteria, algae, phytoplankton, and fungi, become key components of sponges' natural diets during the filter feeding process. In addition, sponges also have diversified microbiome that accounts for up to 40% of the sponge biomass. Sponge-microbe symbioses are considered to promote sponges by providing sustenance, transporting waste products or active metabolites, chemical defense and contribute to mechanical structure (in general) [5].

Microorganisms found in marine animals have a huge potential as a source of bioactive compounds [6]. Sponge relationships are important for exploring biologically active substances that can be used to develop pharmaceuticals, agrochemicals, and biochemical reagents, as well as their lead molecules. It is hypothesized that symbiotic marine microorganisms harbored by sponges are the original producers of these bioactive compounds [7].

Marine fungi belong to a diversity of families; however, they appear to be in low quantities in seawater (in relative to bacteria) and contribute for only 0.6% of the global fungal diversity. The definition of a marine fungus is broad and based on the habitat [8]. Obligate marine fungi are those that grow and sporulate exclusively in a marine or estuary ecosystem; facultative marine fungi are those that can grow and sporulate including both freshwater and terrestrial ecosystems.

The potential fungal origin of a mitochondrial intron presents in the sponge Tetilla sp., which was thought to have emerged from a cross-kingdom horizontal gene transfer, has also been viewed as indirect evidence for a symbiotic relationship between fungus and a sponge [7, 9]. Indirect evidence of interactions between marine sponges and fungi was also provided by the detection of fungal introns in the genomes of some marine sponge species that were most probably acquired by horizontal gene transfer [9].

Marine sponges provide another habitat for fungi [5, 10], knowledge of sponge-associated fungal diversity remains scarce [11]. Marine fungi have provided a major source of new biological natural products, because of their characteristic properties with reference to temperature, nutrients, competition, and salinity [12]. Fungi have been repeatedly isolated from many sponge species [13, 14]. An extensive survey also revealed that there are thousands of other fungi-derived bioactive metabolite families that are yet to be known [15]. Sponge-associated fungi have been reported to create structurally distinct bioactive compounds compared with terrestrial [16, 17].

Considering the fact that researchers do not know much about the fungal life cycle in sponges and other environmental fungi (Richards et al., 2012), it is fascinating to hypothesis about the role of sponge-associated fungi. Many sponge-derived fungi have been found to produce bioactive substances, indicating that they may be involved in chemical host protection [18].

Thirunavukkarasu et al. [19] investigated the filamentous fungal symbionts of 10 marine sponge species from Rameswaram, southern India. The findings indicate that fungal symbionts of marine sponges are extremely diverse. Acremonium, Alternaria, Aspergillus, Cladosporium, Fusarium, and Penicillium were frequently isolated. A few fungi produced acetylcholinesterase inhibitors. Fungi associated with marine sponges have been investigated more avidly for their potential technological applications owing to their ability to synthesize metabolites of novel molecular architectures and bioactivities [20–23].

Several Antarctic sponges of the phylum Ascomycota were a rich source of associate fungi and novel bioactive compounds, with some of them having antibacterial, antitumoral, and antioxidant potential, according to a study on the fungal community using a culture-dependent technique [24].

Many findings proved that sponge-derived fungi are the true biosynthetic origin, able to produce secondary metabolites such as jasplakinolide [25]. Number of fungal strains from marine sponges have been isolated and belong to three phyla, namely Ascomycota, Zygomycota, and mitosporic fungi [20, 26, 27].

1.1 Cultivation of sponge-associated fungi

Sponge tissues are sensitive, and using harsh surface sterilization techniques to isolate their endosymbionts causes sponge disintegration resulting in symbiont death [19]. Most of the surface contaminants can be removed with either a milder sterilization using 70% ethanol for 30 seconds and then washing with seawater or a thorough washing of sponge tissue segments in sterile seawater [28]. Fast-growing fungi, such as *Aspergillus* or *Penicillium*, grow in these conditions, whereas slow-growing species, if present, may go undetected. Weeding out fast-growing fungi or improving the isolation medium with Rose Bengal, with an antibiotic as in isolation techniques for endophytic fungi [10].

In traditional plating method, one gram of sponge sample was mixed in 9 and 99-ml sterile water blank, respectively. This suspension was serially diluted up to 10^{-4}.· Diluted sample was taken from 10^{-3} and 10^{-4} dilutions and was pour plated with 15 ml–20 ml potato dextrose agar (PDA) and incubated at room temperature (28 ± 2°C) for 5 days [29, 30]. Instead, to isolate fungal symbionts, the sponge's water can be squeezed onto nutritive agar medium and cultured [21]. Plate out comminuted sponge tissues on nutritive agar medium as another method of isolating intracellular fungus (J. F. [31]).

Wei et al. [32] applied cultivation-dependent approach to study the fungal diversity in the Hawaiian sponges *Gelliodes fibrosa*, *Haliclona caerulea,* and *Mycale armata* dependent approach. The cultivated fungal isolates belonged to at least 25 genera of Ascomycota and one of Basidiomycota, representing eight taxonomic orders. Cultivated fungal isolates were divided into three groups: sponge-generalists (found in all sponge species), sponge-associates (found in more than one sponge species), and sponge-specialists (found only in one sponge species) [33]. Caballero-George et al. [34] isolated a total of 369 marine sponges that were collected along Panama's coasts in high biodiversity areas. A total of 2263 fungal isolates were recovered from the sponges. Calabon et al. [35] found that Aspergillus was the most dominant genus among 22 genera of ascomycetes isolated from mangrove-originated sponges in Aklan, Philippines, with 23 isolates, followed by Mycelia (n = 21 isolates) and Penicillium (n = 14 isolates).

Marine fungi are especially adept at living on or inside other living organisms such as sponges, corals algae, and even other fungi [36]. Unique metabolic pathways have evolved in halotolerant marine fungal species that are responsive to salt concentrations. Fungi must have osmoregulatory mechanisms that signal the synthesis of polyols and amino compounds while also increasing the concentration of cytoplasmic ions in order to develop in the marine environment. Because the biosynthesis of these osmoregulating solutes is energy-intensive, fungus may release least secondary metabolites or produce them at a slower rate while exposed to high salinity levels [37].

The Endolithic fungus genus Koralionastes is always found in close association with encrusting sponges, which is an interesting observation. Future research on the relationships between fungi and their marine hosts will help our understanding of the marine ecosystem, and it may lead to improved collecting methods and the isolation of chemically unknown species [20].

1.2 Morphological and molecular identification of fungi

Sponge-associated fungi, in particular, have been proven to be the richest source of various bioactive metabolites and novel metabolites. The ecological function and connections of sponge fungi, on the other hand, are mostly unknown. More specific evidence for sponge-associated fungal functions is required. Fungal functions linked with sponges must be provided. The sponge-associated fungal function analysis can employ an activity-based analysis technique, but it will be limited in some cases because the culture condition during the bioassay may not be optimal for the production of linked bioactives.

For natural product exploration, many sponge marine-derived fungi have been isolated by many researchers. It was identified primarily by microscopic observation using wet mount and lacto phenol cotton blue stain preparations. Traditional monographs, polyphasic taxonomic approach, and molecular nucleotide sequences of marker DNA such as ITS, 18 s, and others are used to identify marine fungus strains [24, 38, 39]. Though, ITS rDNA regions are most often used to identify species and strain-level fungal diversity. DNA barcode data for approximately 100,000 fungal isolates were generated using sequences of two nuclear ribosomal genetic markers, the Internal Transcribed Spacer and 5.8S gene (ITS) and the D1/D2 domain of the 26S Large Subunit (LSU).

1.3 Taxonomic databases resources for marine fungi

The internet has become a vital source of information for millions of individuals. Over the last few decades, fungal research has broadened its scope, generating a wealth of information that has led to the establishment of many site dedicated to various aspects of mycology and also exclusively marine fungi such as http://www.marinespecies.org/.and,http://pubs.rsc.org/marinlit.

All genera of fungi, including marine fungi, are classified and details are provided in the database (http:// fungalgenera.org/). The Indian marine fungal database (**Figure 1**) is another resource (www.fungifromindia.com/), which is linked to MycoBank and provides 233 strains of marine fungi identified in India. The World Register of Marine Species (WoRMS) (www.marinespecies.org) aims to provide a comprehensive and definitive listing of all marine life forms' names [40–42]. www.marinefungi.org is a marine web portal. This web portal provides researchers with

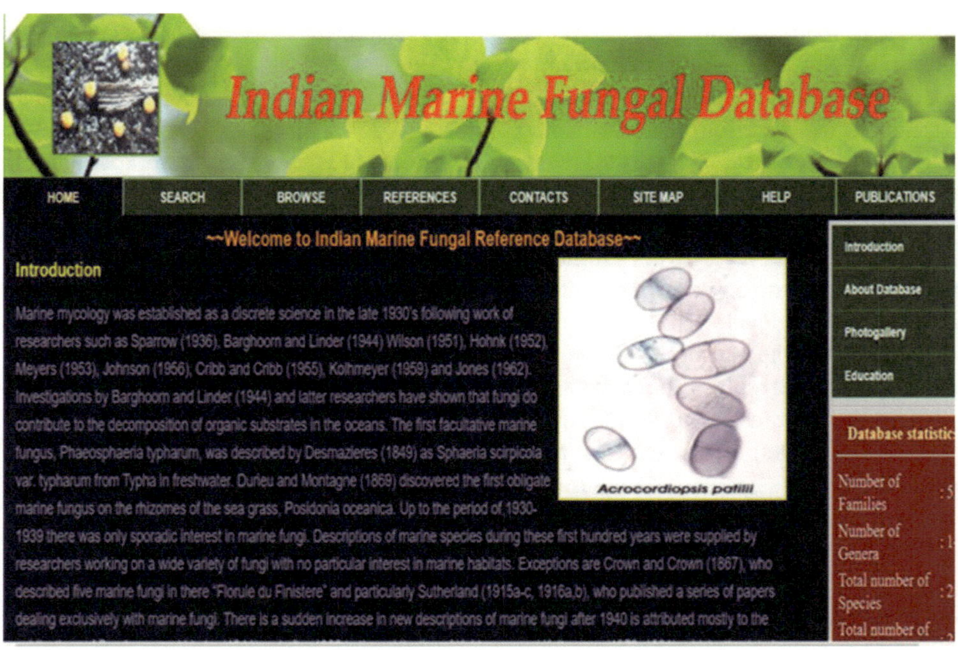

Figure 1.
Indian marine fungi database (www.fungifromindia.com).

access to the classification, detailed descriptions, and worldwide distribution of all known marine and marine-derived fungi [43].

2. Bioactive natural metabolites from sponge-associated fungi

Marine sponge-associated fungi are one such group that has been reported to be a crucial and invaluable source of novel therapeutic agents possessing several bioactive properties including free radical scavenging activity, neuritogenic activity, anticancer activity, and kinase inhibition, etc. The exploration of fungal metabolites has significantly increased after marine fungi, especially sponge-associated fungi have been reported to produce structurally unique bioactive compounds [44, 45].

The first metabolite reported from a sponge-derived fungus was Trichoharzin, which was isolated from a strain of *Trichoderma harzianum* associated with the sponge *Mycale cecilia* in 1993.

2.1 Anticancer compounds

The diversity of biochemical properties of sponges had been demonstrated by the continued discovery of novel compounds, having pharmacological properties [46]. The marine-derived fungus Aspergillus sp., which was obtained from the sponge *Xestospongia testudinaria*, was collected from the South China Sea that gave two phenolic bisabolane sesquiterpenoid dimers, disydonols A and C exhibited in vitro moderate cytotoxicity toward HepG- 2 and Caski human tumor cell lines (IC50 values of 9.31 and 12.40 μg/mL) [47]. Fungi *Stachylidium* spp. was isolated from the sponge *Callyspongia cf. C. flammea*. Chemical investigation of the bioactive fungal extract led to the isolation of the novel phthalimidine derivatives marilines A1 and A2. Both enantiomers, marilines A1 and A2 inhibited human leukocyte elastase (HLE) with an IC50 value of 0.86 μM [48]. The fungal species Aspergillus, which is widespread across the globe, is also the major source of bioactive molecules in marine sponges. The bulk of the 680 fungal strains derived from 16 sponge species around the world are mostly from the genera Aspergillus and Penicillium [49].

Three marine sponges, *Tedania anhelans*, *Myxilla arenaria*, *Callyspongia fibrosa*, were collected from Vizhinjam and Kovalam in Kerala. Aspergillus sp. MCCF 103, Aspergillus sp. MCCF 111, Aspergillus sp. MCCF 114, Penicillium sp. MCCF 115, and Aspergillus sp. MCCF were isolated and identified. These strains have significant cytotoxic activity on NCI-H460 lung cancer cells lines [50]. Yellow-colored compounds 2-(2′, 3-epoxy-1′,3′-heptadienyl)-6-hydroxy-5-(3- methyl-2-butenyl) benzaldehyde and 1,8-dihydroxy-6-methoxy- 3-methyl-9,10-anthracenedione (physcion) are extracted from the marine sponge Mycale sp., associated fungus *Eurotium cristatum* [48].

Violaceimides A and B, two methyl succinimide-based sulfur-bearing compounds, were isolated from the sponge-associated fungal strain *Aspergillus violaceus* WZXY-m64-17.

Both compounds suppressed human leukemia U937 growth with IC50 values of 5.3 ± 0.4 and 1.8 ± 0.6 mM, respectively, as well as human colorectal cancer cell HCT-8 with IC50 values of 1.5 ± 0.28 mM [51]. Mactanamide, a diketopiperazine alkaloid, was isolated from the marine sponge *Stylissa* sp. derived fungus *Aspergillus flocculosus*, which was collected in Vietnam. The isolated compound was screened for antiproliferation activity, and it proved a significant effect of non-cytotoxic suppression on osteoclast differentiation (Shin et al., 2017).

Preussin, a hydroxyl pyrrolidine derivative, was isolated from *Aspergillus candidus* KUFA 0062, a fungus associated to sea sponges. The antiproliferative and cytotoxic activities of this pyrrolidine derivative have been tested in breast cancer cells (SKBR3, MCF7, and MDA-MB-231), as well as MCF12A, a non-tumor cell line. Various assays have been used to examine cell morphology for ki67 and caspase-3, as well as 3D (multicellular aggregates) and 2D (monolayer) culturing tests. Preussin-exposed cells morphological study indicated apoptosis, which was confirmed by caspase-3 immunohistochemistry. 3D culture cells were less sensitive, and preussin-exposed cells morphological analysis revealed apoptosis, which was confirmed by caspase-3 immunohistochemistry [52].

Bioactive component methyl averantin is produced by *Aspergillus versicolor* in association with the sponge *Petrosia* sp. This secondary metabolite belongs to the anthraquinone family. Methyl averantin has a high cytotoxic activity, with an IC50 .4–1.1 μg/ml in cancer cell lines such as A-549, HCT-15, SK-MEL-2, SK-OV-3, and XF- 498 [53]. The compounds heterocornols AC, FH, methyl(2formyl3hydroxyphenyl) propanoate, agropyrenol, and vaccinol G have been isolated from the fungus *Pestalotiopsis hetero cornis* XWS03F09 associated with the marine sponge *Phakellia fusca* and have cytotoxicity against four human cancer cell lines and antimicrobial activity [54]. The Asteltoxins E and F polyketides were isolated from the marine sponge-derived fungus Aspergillus sp. SCSIO XWS02F40. With IC50 values of 6.2 ± 0.08 and 8.9 ± 0.3 mM, respectively, asteltoxin E and F demonstrated potent antiviral activity against influenza virus A subtype H3N2 (A/H3N2). Furthermore, asteltoxin E reported to inhibit the activity of influenza virus A subtype H1N1 (A/) [55].

The fungus *Arthrinium arundinis* ZSDS1-F3 was isolated from the marine sponge *Phakellia fusca* in the Xisha Islands of China, from that cytochalasin K was extracted that showed cytotoxicity against K562, A549, Huh-7, H1975, MCF-7, U937, BGC823, HL60, HeLa, and MOLT-4 cell lines, with IC50 values of 10.5, 13.7, 10.9, 19.1, 11.1, 47.4, and 11.8 μM respectively [56].

Elissawy et al. (2017) extracted Curvularin, Cyclo(L- Pro-L-Ile), and Cyclo(L-Tyr-L-Pro), from the fungus *Aspergillus versicolor* isolated from the black sponge *Spongia officinalis*, which play inhibitory activity against HCV NS3/4A protease.

Therapeutic enzymes are used to treat diseases such as cancer, severe disorders such as autism, chronic lung disease, and multiple sclerosis, although cancer seems to be the most potential therapeutic application for enzymes. Therapeutic enzymes, it seems out, have a unique ability to facilitate high-affinity interactions with unrelated cancer-related proteins. Endophytic fungi recovered from the marine soft sponge *Aplysina fistularis*, produce L-asparaginase [57].

2.2 Antimicrobial compounds

Polyketide-derived alkaloids, terpenes, peptides, and combined biosynthetic chemicals are prominent classes of secondary metabolites produced by marine sponge-derived fungus. Miriam et al. [58] isolated several bioactive secondary metabolites from the fungi *P. raistrikii* associated *Axinella* cf. corrugate (sponge), including (4-methoxy-5-3-methoxybut-1-enyl)-6- methyl-2H- pyran-2-one, a new metabolite isolated from the *Penicillium paxilli* strain MaGK, Norlique xanthone, also known as 1, 3,6- trihydroxy-8-methyl-9H-xanthen-9 [32].

Triazolic compound was extracted from *Aspergillus clavatus* MFD15 that is associated with marine sponges. This compound is found to 50% inhibit *E. coli, S. aureus*, and *S. epidermidis* [59].

Fugal extract of *Aspergillus sydowii* from the waters of Riung, East Nusa Tenggara, Indonesia, was associated with marine sponge *Axinella* sp. and showed antibacterial activity. These extracts have bioactivity against E. coli and S. aureus. The maximum zone is obtained from MG KN-15-3-1-3 extract, with inhibition zones of 10.71 mm and 10.98 mm against *E. coli* and *S. aureus* [60]. Likewise, Austalide U, a meroterpenoid, has been produced by the sponge-derived fungus *Aspergillus aureolatus* HDN14-107. It showed antiviral efficiency against A/H1N1 virus [55].

Marine fungi are well known for producing a wide range of secondary metabolites, including numerous life-saving therapeutics [61]. MDR *Escherichia coli* has been linked to a variety of infectious diseases, as well as urinary tract infection, nosocomial bloodstream infection, meningitis, bacteremia, and gastrointestinal disease. There were 29 marine sponge-associated fungi isolated from nine sponges. Among 29 sponge-associated fungi screened, there were seven isolates that showed antibacterial activity against MDR *E. coli* [62]. Sponge-associated *Aspergillus sp* LS116 produced aspergill steroid A, a C23 steroid with a bicycloA/B ring (With an MIC value of 16 mg/mL1, this compound showed strong antibacterial activity against *V. harveyi*, indicating that aspergillsteroid A.) It could be considered one of the promising agents for aquatic disease control in the future Guo et al. [63].

Daldinia eschscholtzii is a fungus isolated from an Indonesian sponge called *Xestospongia* sp. located in Karimunjawa National Park in Central Java, Indonesia. Karimanone is a novel chromanone-type compound found and characterized from *D. eschscholtzii*, and it has three biosynthetically related metabolites. With an MIC of 62.5 g/ml for compound 2 and 125 g/ml for compounds 1, 3, and 4, all of the compounds were effective against a multidrug-resistant strain of *Salmonella enterica ser. Typhi*. [64].

Two cyclic tetrapeptides, sartory glabramides A and B and a bis-indoly lmethyl diketopiperazine, fell utanine A epoxide together with aszonalenin (3R)-3-(1H-indol-3-ylmethyl)-3,4-dihydro-1H-1,4-benzodiazepine-2,5-dione takakiamide), (11aR)-2,3-dihydro-1H-pyrrolo benzodiazepine- 5,11(10H,11aH)-dione and fellutanine A were isolated from the marine-derived fungus *N. glabra* KUFA 0702, which was isolated from the marine sponge *Mycale* sp., collected from the coral reef at Samaesarn Island, Thailand. The antibacterial activity of all identified compounds was tested against two bacterial pathogens, *Staphylococcus aureus* ATCC 46645 and *E. coli* ATCC 25922, as well as three fungal isolates, *Aspergillus fumigatus* ATCC 46645, *Trichophyton rubrum* ATCC FF5, and *Candida albicans* ATCC 10231 [65].

Chu et al. [66] isolated and identified the *Aspergillus versicolor* strain TS08 associated with South China Sea sponge *Holoxea* sp. also extracted the cyclo (L-Trp-L-Phe). The highest yield of cyclo (L-Trp-L-Phe), 13.24 mg/g (per crude extract of EtOAc), 2.51% of cell dry weigh, was obtained on the tenth day of the fungal cultivation. Scopel et al. [67] separated Arvoredol from Penicillium sp. F37, which was isolated from the marine sponge *Axinella corrugate*.

It is a chlorinated polyketide that contains 6,7-dihydro-4(5H) benzofuranone. Arvoredol inhibited biofilm formation by the human pathogen *S. epidermidis* by 40% at a concentration of 125 μgmL−1 by 40%.

The extract of *Penicillium chrysogenum*, obtained from the marine sponge *Tedania anhelans*, showed antimycobacterial activity [68], another sponge-associated Penicillium sp. produced citrinin, which has antibacterial and cytotoxic properties [69, 70]. Sabdaningsih et al. [71] also isolated *P. citrinum* WK-P9, a sponge-associated fungus, having been used to produce citrinin derivatives. Penicitrinol J. It was characterized, revealing a monomer connection previously unknown in citrinin derivatives. It inhibits the growth of *B. subtilis* JH642, *B. megaterium* DSM32, and *M. smegmatis* ATCC607.

On the one side, the extensive use of antibiotics has increased the prevalence of antibiotic resistance; on the other side, the use of immunosuppressive agents after transplantation has significantly increased the incidence of fungal infections. Phoma sp., a sponge-derived fungus, provides unique lactone compound capable of inhibiting several human pathogens such as *C. albicans* and *Aspergillus fumigatus* [18]. *Curvularia lunata*, a fungus, was isolated from the sponge *Cinachyrella australiensis* from the Karimunjawa Islands in Indonesia. *C. lunata* fungal extract demonstrated promising antibacterial activity against MDR *S. pneumoniae* [72].

Cytotoxic polyketides compounds were extracted from the sponge-derived fungus *Aspergillus versicolor*. Bioactivity-guided fractionation was used to isolate a new peptide in a subsequent investigation. Approximately 20 peptides have been reported from sponge-derived fungi, including efrapeptins Eα, H, RHM3, RHM4 (from two fungi, *Acremonium* sp. and *Metarrhizium* sp),4 homodestcardin (from *Fusarium graminearum*),5 clonostachysins A and B (from *Clonostachys rogersoniana*),6 a cyclodepsipeptide (from a Clonostachys sp),7 linear octapeptides (from an Acremonium sp),8 petrosifungins A and B (from *Penicillium brevicompactum*),9 fellutamides A and B (from *Penicillium fellutanum*),10 and halovirus (from a Scytalidium sp). Fungi *Penicillium chrysogenum* and *Stachybotrys chartarum* derived from sponge have been found to inhibit at different stages of the HIV viral cycle [73].

2.3 Antifungal activity

Peniciadametizine A and B were extracted from *Penicillium adamatzioides*, the marine sponge and have antifungal activity against *Alternaria brassicae* [13].*Penicillium cf. montanense*, a marine fungus that produces xestolactone B, is an associated fungus of the marine sponge *Xetospongi aexigua*. This compound is antifungal against *C. albicans* [74].

Sixty-seven sponges were collected from four different areas of Indonesian water. For screening the active isolates, an antagonistic test was performed against Malassezia furfur, Trichophyton sp., and C. albicans using the cross-streak method. Lampung Bay, Seribu Islands, Karimunjawa Islands, and Wakatobi Island had sponge-isolates ratios of 106%, 90% 210%, and 115%, respectively. The sponge collected from the Wakatobi Islands has one of the most active isolates against *M. furfur*, *Trichophyton* sp., and *C. albicans* [71].

A number of compounds synthesized by fungal symbionts from sponges may have agricultural implications. *Penicillium adametzioides* AS-53 produces peniciadametizine A, which is particularly active against the plantpathogenic fungus *Alternaria brassicae* (MIC 4.0 g/mL) [75]. Fungi isolated from sponges proved to be bioremediation agents, for instance by degrading the pesticide DDD (1,1-dichloro-2,2-bis-(4-chlorophenyl)ethane), nowadays banned but still persistent in the environment [76].

3. Conclusion and future prospective

Marine environment represents an untapped source of fungal diversity, in comparison to sponge-associated prokaryotic microorganisms; there are few reports on the diversification of sponge-associated fungi. Marine sponge-associated fungi are rich in metabolites, which are less understood. Fungi associated with sponges are the most potent source of new natural compounds and display diverse biological activities. Based on recent research studies, marine fungal metabolites will find application toward pharmaceuticals, cosmeceuticals, nutraceuticals, etc.; from this review, it is also important to remember that secondary metabolite profiles differ from the same fungus species originating from various sponge species. As a conclusion, a systematic search for fungus and fungal metabolites in sponge species from various geographical regions is an essential step in bioprospecting.

Author details

Vasanthabharathi Venkataraman*, Kalaiselvi Vaithi and Jayalakshmi Singaram
Faculty of Marine Sciences, CAS in Marine Biology, Annamalai University, Parangipettai, Tamil Nadu, India

*Address all correspondence to: vasanthab2505@gmail.com

References

[1] Haefner B. Drugs from the deep: Marine natural products as drug candidates. Drug Discovery Today. 15 Jun 2003;**8**(12):536-544. DOI: 10.1016/s1359-6446(03)02713-2. PMID: 12821301

[2] Mora C, Tittensor DP, Adl S, Simpson AGB, Worm B. How many species are there on earth and in the ocean? PLoS Biology. 2011;**9**(8):1-8. DOI: 10.1371/journal.pbio.1001127

[3] Mu WEG, Mu IM. Origin of the metazoan immune system: Identification of the molecules and their functions in sponges. 2003;**1. 43**:281-292. Available from: https://academic.oup.com/icb/article/43/2/281/609530

[4] Fieseler L, Horn M, Wagner M, Hentschel U. Fieseler_04_sponge.pdf. Applied and Environmental Microbiology. 2004;**70**(6):3724-3732. DOI: 10.1128/AEM.70.6.3724

[5] Simister RL, Deines P, Botté ES, Webster NS, Taylor MW. Sponge-specific clusters revisited: A comprehensive phylogeny of sponge-associated microorganisms. Environmental Microbiology. 2012;**14**(2):517-524. DOI: 10.1111/j.1462-2920.2011.02664.x

[6] Satheesh S, Ba-Akdah MA, Al-Sofyani AA. Natural antifouling compound production by microbes associated with marine macroorganisms — A review. Electronic Journal of Biotechnology. 2016;**21**:26-35. DOI: 10.1016/j.ejbt.2016.02.002

[7] Proksch P, Edrada RA, Ebel R. Drugs from the seas - Current status and microbiological implications. Applied Microbiology and Biotechnology. 2002;**59**(2-3):125-134. DOI: 10.1007/s00253-002-1006-8

[8] Kohlmeyer J, Volkmann-Kohlmeyer B. Fungi from coral reefs: A commentary. Mycological Research. 2003;**107**(4):386-387. DOI: 10.1017/S0953756203227775

[9] Rot C, Goldfarb I, Ilan M, Huchon D. Putative cross-kingdom horizontal gene transfer in sponge (Porifera) mitochondria. BMC Evolutionary Biology. 2006;**6**:1-11. DOI: 10.1186/1471-2148-6-71

[10] Taylor MW, Radax R, Steger D, Wagner M. Sponge-associated microorganisms: Evolution, ecology, and biotechnological potential. Microbiology and Molecular Biology Reviews. 2007;**71**(2):295-347. DOI: 10.1128/MMBR.00040-06

[11] Webster NS, Taylor MW. Marine sponges and their microbial symbionts: love and other relationships. Environmental Microbiology. 2011. (In press) DOI: 10.1111/j.1462-2920.2011.02460.x

[12] Swathi J. Marine fungal metabolites as a rich source of bioactive compounds. African Journal of Biochemistry Research. 2013;**7**(10):184-196. DOI: 10.5897/ajbr12.068

[13] Liu Y, Mándi A, Li XM, Meng LH, Kurtán T, Wang BG. Peniciadametizine A, a dithiodiketopiperazine with a unique spiro[furan-2,7′-pyrazino[1,2-b] [1,2] oxazine] skeleton, and a related analogue, peniciadametizine B, from the marine sponge-derived fungus Penicillium adametzioides. Marine Drugs. 2015;**13**(6):3640-3652. DOI: 10.3390/md13063640

[14] Paz Z, Burdman S, Gerson U, Sztejnberg A. Antagonistic effects of the endophytic fungus Meira geulakonigii on the citrus rust mite Phyllocoptruta oleivora. Journal of Applied Microbiology. 2007;**103**(6):2570-2579. DOI: 10.1111/j.1365-2672.2007.03512

[15] Debbab A, Aly AH, Lin WH, Proksch P. Bioactive Compounds from Marine Bacteria and Fungi. 2010;**3**(5):544-563. DOI: 10.1111/j.1751-7915.2010.00179.x

[16] Hasan S, Ansari MI, Ahmad A, Mishra M. Major bioactive metabolites from marine fungi: A review. Bioinformation. 2015;**11**(4). Available from: www.bioinformation.net

[17] Sithranga Boopathy N, Kathiresan K. Anticancer drugs from marine flora: An overview. Journal of Oncology. 2010;**2010**. DOI: 10.1155/2010/214186

[18] Indraningrat AA, Smidt H, Sipkema D. Bioprospecting sponge-associated microbes for antimicrobial compounds. Marine Drugs. 2016;**14**(5):87. DOI: 10.3390/md14050087

[19] Thirunavukkarasu N, Suryanarayanan TS, Girivasan KP, Venkatachalam A, Geetha V, Ravishankar JP, et al. Fungal symbionts of marine sponges from Rameswaram, southern India: Species composition and bioactive metabolites. Fungal Diversity. 2012;55(July):37-46. DOI: 10.1007/s13225-011-0137-6

[20] Bugni TS, Ireland CM. Marine-derived fungi: A chemically and biologically diverse group of microorganisms. Natural Product Reports. 2004;**21**(1):143-163. DOI: 10.1039/b301926h

[21] Kjer J, Debbab A, Aly AH, Proksch P. Methods for isolation of marine-derived endophytic fungi and their bioactive secondary products. Nature Protocols. 2010;**5**(3):479-490. DOI: 10.1038/nprot.2009.233

[22] Masi M, Evidente A. Fungal bioactive anthraquinones and analogues. Toxins. 2020;**12**(11). DOI: 10.3390/toxins12110714

[23] Palis FG, Florencia G. Research to impact : Case studies for natural resource management for irrigated rice in Asia. 2010 (March 2015)

[24] Henríquez M, Vergara K, Norambuena J, Beiza A, Maza F, Ubilla P, et al. Diversity of cultivable fungi associated with Antarctic marine sponges and screening for their antimicrobial, antitumoral and antioxidant potential. World Journal of Microbiology and Biotechnology. 2014;**30**(1):65-76. DOI: 10.1007/s11274-013-1418-x

[25] Jensen PR, Fenical W. Secondary metabolites from marine fungi In: Hyde KD, editor. Fungi in Marine Environments. Hong kong: Fungal Diversity Press; 2002. pp. 293-315

[26] Gesa HW, Gabrielle M, Draeger KS, Aust H-J, Schulz B. Fungi from marine sponges: Diversity, biological activity and secondary metabolites. Mycological Research. 2000;**104**(11):1354-1365. DOI: 10.1017/S0953756200003117

[27] Morrison-Gardiner S. Dominant fungi from Australian coral reefs. Fungal Divers. 2002;**9**:105-121

[28] Ebada SS, Schulz B, Wray V, Totzke F, Kubbutat MHG, Müller WEG, et al. Arthrinins A–D: Novel diterpenoids and further constituents from the sponge derived fungus Arthrinium sp. Bioorganic & Medicinal Chemistry. 2011;**19**(15):4644-4651. DOI: 10.1016/J.BMC.2011.06.013

[29] Meenupriya J, Thangaraj M. Isolation and molecular characterization of bioactive secondary metabolites from Callyspongia spp. associated fungi. Asian Pacific Journal of Tropical Medicine. 2010;**3**(9):738-740. DOI: 10.1016/S1995-7645(10)60177-0

[30] Vasanthabharathi V. Bioactive potential of symbiotic bacteria and fungi from marine sponges. African Journal of Biotechnology.

2012;**11**(29):7500-7511. DOI: 10.5897/ajb11.1378

[31] Wang JF, Lin XP, Qin C, Liao SR, Wan JT, Zhang TY, et al. Antimicrobial and antiviral sesquiterpenoids from sponge-associated fungus, Aspergillus sydowii ZSDS1-F6. Journal of Antibiotics. 2014;**67**(8):581-583. DOI: 10.1038/ja.2014.39

[32] Wei M, Chen G, Wang Y, Zhang X. Of chrodrimanin B from a marine fungus Aspergillus sp. 2011;**47**(4):506-508

[33] Manwaring HR, Bligh HFJ, Yadav R. The challenges and opportunities associated with biofortificati of pearl millet (Pennisetum glaucum) with elevated levels of grain iron and zinc. Frontiers in Plant Science. 2016;7(DECEMBER2016):1-15. DOI: 10.3389/fpls.2016.01944

[34] Caballero-George C, Bolaños J, Ochoa E, Carballo JL, Cruz JA, Elizabeth Arnold A. Protocol to isolate sponge-associated fungi from tropical waters and an examination of their cardioprotective potential. Current Trends in Biotechnology and Pharmacy. 2010;**4**(4):881-899

[35] Calabon MS, Sadaba RB, Campos WL. Fungal diversity of mangrove-associated sponges from New Washington, Aklan, Philippines. Mycology An International Journal on Fungal Biology. 2018. DOI: 10.1080/21501203.2018.1518934

[36] Kiran GS, Sekar S, Ramasamy P, Thinesh T, Hassan S, Lipton AN, et al. Marine sponge microbial association: Towards disclosing unique symbiotic interactions. Marine Environmental Research. 2018;**140**:169-179. DOI: 10.1016/J.MARENVRES.2018.04.017

[37] Keller NP. Fungal secondary metabolism: Regulation, function and drug discovery. Nature Reviews Microbiology. 2019;**17**(3):167-180. DOI: 10.1038/s41579-018-0121-1

[38] Alwakeel SS. Molecular identification of fungi isolated from coastal regions of Red Sea. Journal of the Association of Arab Universities for Basic and Applied Sciences. 2017;**24**(1):115-119. DOI: 10.1016/j.jaubas.2016.10.001

[39] Bovio E. Marine fungi from sponges: biodiversity, chemodiversity and biotechnological applications To cite this version: HAL Id: tel-02514804 biotechnological applications Elena Bovio. 2020. pp. 36-41

[40] Bhat R, Sridhar KR. Marine fungi and novel metabolites. Frontiers in Fungal Ecology, Diversity and Metabolites. 2009:51. DOI: 10.13140/RG.2.1.1039.0167

[41] Hyde KD, Dong Y, Phookamsak R, Jeewon R, Bhat DJ, Jones EBG, et al. Fungal diversity notes 1151-1276: Taxonomic and phylogenetic contributions on genera and species of fungal taxa. Fungal Diversity. 2020;**100**(1). DOI: 10.1007/s13225-020-00439-5

[42] Wijayawardene NN, Hyde KD, Tibpromma S, Wanasinghe DN, Thambugala KM, Tian Q, et al. Towards incorporating asexual fungi in a natural classification: Checklist and notes 2012-2016. Mycosphere. 2017;**8**(9):1457-1555. DOI: 10.5943/mycosphere/8/9/10

[43] Jones EBG, Pang KL, Abdel-Wahab MA, Scholz B, Hyde KD, Boekhout T, et al. An online resource for marine fungi. Fungal Diversity. 2019;**96**(1). DOI: 10.1007/s13225-019-00426-5

[44] Imhoff JF, Labes A, Wiese J. Bio-mining the microbial treasures of the ocean: New natural products. Biotechnology Advances. 2011;**29**(5):468-482. DOI: 10.1016/J.BIOTECHADV.2011.03.001

[45] Suryanarayanan TS. The diversity and importance of fungi associated with marine sponges. Botanica Marina. 2012;**55**(6):553-564. DOI: 10.1515/bot-2011-0086

[46] Ridley CP, John Faulkner D, Haygood MG. Investigation of Oscillatoria spongeliae-dominated bacterial communities in four dictyoceratid sponges. Applied and Environmental Microbiology. 2005;**71**(11):7366-7375. DOI: 10.1128/AEM.71.11.7366-7375.2005

[47] Sun LL, Shao CL, Chen JF, Guo ZY, Fu XM, Chen M, et al. New bisabolane sesquiterpenoids from a marine-derived fungus Aspergillus sp. isolated from the sponge Xestospongia testudinaria. Bioorganic & Medicinal Chemistry Letters. 2012;**22**(3):1326-1329. DOI: 10.1016/j.bmcl.2011.12.083

[48] Almeida AP, Dethoup T, Singburaudom N, Lima R, Vasconcelos MH, Pinto M, et al. The *in vitro* anticancer activity of the crude extract of the sponge-associated fungus Eurotium cristatum and its secondary metabolites. Journal of Natural Pharmaceutical. 2010;**1**:25-29. DOI: 10.4103/2229-5119.7358331

[49] Nguyen MTHD, Thomas T. Diversity, host-specificity and stability of sponge-associated fungal communities of co-occurring sponges. PeerJ. 2018;**2018**(6):1-26. DOI: 10.7717/peerj.4965

[50] Lekshmi N, Umar, Dhaneesha M, Joseph R, Ravinesh R, Sajeevan TP. Endophytic fungi isolated from the marine sponges as a source of potential bioactive compounds. Journal of Aquatic Biology & Fisheries. 2020;**8**:58-66

[51] Yin J, Zhang C, Huang J, Zhang J, Liu D, Huang J, et al. Violaceimides A–E, sulfur-containing metabolites from a sponge-associated fungus Aspergillus violaceus. Tetrahedron Letters. 2018;**59**(33):3157-3160. DOI: 10.1016/J.TETLET.2018.05.085

[52] Zhou G, Chen X, Zhang X, Che Q, Zhang G, Zhu T, et al. Supporting Information Prenylated p -Terphenyls from a Mangrove Endophytic Fungus Aspergillus candidus LDJ-5 Open Studio for Druggability Research of Marine Natural Products, Pilot National. 2020

[53] Lee J, Kim H, Yu H, Chung JY, Oh CH, Yoo KH, et al. Discovery and initial SAR of pyrimidin-4-yl-1H-imidazole derivatives with antiproliferative activity against melanoma cell lines. Bioorganic & Medicinal Chemistry Letters. 2010;**20**(5):1573-1577. DOI: 10.1016/j.bmcl.2010.01.064

[54] Lei H, Lin X, Han L, Ma J, Dong K, Wang X, et al. Polyketide derivatives from a marine-sponge-associated fungus Pestalotiopsis heterocornis. Phytochemistry. 2017;**142**:51-59. DOI: 10.1016/J.PHYTOCHEM.2017.06.009

[55] Tian YQ, Lin XP, Wang Z, et al. Asteltoxins with antiviral activities from the marine sponge-derived fungus Aspergillus sp. SCSIO XWS02F40. Molecules. 2015;**21**(1):E34. Published 2015 Dec 26. DOI: 10.3390/molecules21010034

[56] Eamvijarn A, Gomes NM, Dethoup T, Buaruang J, Manoch L, Silva A, et al. Bioactive meroditerpenes and indole alkaloids from the soil fungus Neosartorya fischeri (KUFC 6344), and the marine-derived fungi Neosartorya laciniosa (KUFC 7896) and Neosartorya tsunodae (KUFC 9213). Tetrahedron. 2013;**69**(40):8583-8591. DOI: 10.1016/j.tet.2013.07.078

[57] Abbas Ahmed MM. Production, purification and characterization of L-asparaginase from marine endophytic aspergillus sp. ALAA-2000 under

submerged and solid state fermentation. Journal of Microbial & Biochemical Technology. 2015;**07**(03):165-172. DOI: 10.4172/1948-5948.1000199

[58] Kossuga MH, Romminger S, Xavier C, et al. Evaluating methods for the isolation of marine-derived fungal strains and production of bioactive secondary metabolites. Revista Brasileira de Farmacognosia. 2012;**22**(2):257-267. DOI: 10.1590/S0102-695X2011005000222

[59] Manilal A, Sabarathnam B, Kiran GS, Sujith S, Shakir C, Selvin J. Antagonistic potentials of marine sponge associated fungi Aspergillus clavatus MFD15. Asian Journal of Medical Sciences. 2010;**2**(4):195-200. Available from: http://maxwellsci.com/print/ajms/v2-195-200.pdf

[60] Widyaningsih, Trianto, Radjasa, Wittriansyah. IOP Conference Series: Earth and Environmental Science. 2018;**116**:012084

[61] Wang Y-T, Xue Y-R, Liu C-H. marine drugs a brief review of bioactive metabolites derived from deep-sea fungi. Marine Drugs. 2015;**13**:4594-4616. DOI: 10.3390/md13084594

[62] Sibero MT, Radjasa OK, Sabdono A, Trianto A, Triningsih DW, Hutagaol ID. Antibacterial activity of indonesian sponge associated fungi against clinical pathogenic multidrug resistant bacteria. Journal of Applied Pharmaceutical Science. 2018;**8**(2):088-094. DOI: 10.7324/JAPS.2018.8214

[63] Guo L, Zhang F, Wang X, et al. Antibacterial activity and action mechanism of questin from marine *Aspergillus flavipes* HN4-13 against aquatic pathogen *Vibrio harveyi*. 3 Biotech. 2019;**9**(1):14. DOI: 10.1007/s13205-018-1535-1

[64] Sibero MT, Zhou T, Igarashi Y, Radjasa OK, Sabdono A, Trianto A, et al. Chromanone-type compounds from marine sponge-derived Daldinia eschscholtzii KJMT FP 4.1. Journal of Applied Pharmaceutical Science. 2020;**10**(1):001-007

[65] Zin WWM, Buttachon S, Buaruang J, Gales L, Pereira JA, Pinto MMM, et al. A new meroditerpene and a new tryptoquivaline analog from the algicolous fungus Neosartorya takakii KUFC 7898. Marine Drugs. 2015;**13**(6):3776-3790. DOI: 10.3390/md13063776

[66] Chu D, Peng C, Ding B, Liu F, Zhang F, Lin H, et al. Biological active metabolite cyclo (l-Trp-l-Phe) produced by South China Sea sponge Holoxea sp. associated fungus Aspergillus versicolor strain TS08. Bioprocess and Biosystems Engineering. 2011;**34**(2):223-229. DOI: 10.1007/s00449-010-0464-0

[67] Marina S, Mothes B, Lerner CB, Henriques AT, Macedo AJ, Abraham W-R. Arvoredol—An unusual chlorinated and biofilm inhibiting polyketide from a marine Penicillium sp. of the Brazilian coast. Phytochemistry Letters. 2017;**20**:73-76. DOI: 10.1016/j.phytol.2017.04.010

[68] Visamsetti A, Ramachandran SS, Kandasamy D. Penicillium chrysogenum DSOA associated with marine sponge (Tedania anhelans) exhibit antimycobacterial activity. Microbiological Research. 2016;**185**:55-60. DOI: 10.1016/J.MICRES.2015.11.001

[69] Subramani R, Kumar R, Prasad P, Aalbersberg W. Cytotoxic and antibacterial substances against multi-drug resistant pathogens from marine sponge symbiont: Citrinin, a secondary metabolite of Penicillium sp. Asian Pacific Journal of Tropical Biomedicine. 2013;**3**(4):291-296. DOI: 10.1016/S2221-1691(13)60065-9

[70] Vasanthabharathi V, Jayalakshmi S. Bioactive potential of Callyspongia

diffusa associated *Pseudomonus fluorescens* BCPBMS-1 and *Penicillum citrinum*. Microbial Bioactives. 2018;**1**(1):008-013. DOI: 10.25163/microbbioacts.11002A2221300318

[71] Sabdaningsih A, Liu Y, Mettal U, Heep J, Marner M, Radjasa OK, et al. A New Citrinin Derivative from the Indonesian. 2020

[72] Cristianawati O, Sabdaningsih A, Becking LE, Khoeri MM, Nuryadi H, Sabdono A, et al. Biological activity of sponge-associated fungi from Karimunjawa islands, Indonesia against pathogenic streptococcus pneumoniae. Biodiversitas. 2019;**20**(8):2143-2150. DOI: 10.13057/biodiv/d200807

[73] Abdelmohsen UR, Balasubramanian S, Oelschlaeger TA, Grkovic T, Pham NB, Quinn RJ, et al. Potential of marine natural products against drug-resistant fungal, viral, and parasitic infections. The Lancet Infectious Diseases. 2017;**17**(2):e30-e41. DOI: 10.1016/S1473-3099(16)30323-1

[74] Edrada RA, Heubes M, Brauers G, Wray V, Berg A, Gräfe U, et al. Online analysis of xestodecalactones A-C, novel bioactive metabolites from the fungus Penicillium cf. montanense and their subsequent isolation from the sponge Xestospongia exigua. Journal of Natural Products. 2002;**65**(11):1598-1604. DOI: 10.1021/np020085c

[75] El-Hossary EM, Cheng C, Hamed MM, El-Sayed Hamed AN, Ohlsen K, Hentschel U, et al. Antifungal potential of marine natural products. European Journal of Medicinal Chemistry. 2017;**126**:631. DOI: 10.1016/j.ejmech.2016.11.022

[76] Rotter A, Barbier M, Bertoni F, Bones AM, Cancela ML, Carlsson J, et al. The essentials of marine biotechnology. Frontiers in Marine Science. 2021;**8**(March):1-53. DOI: 10.3389/fmars.2021.629629